Quantum Queries
Dr. Anab Whitehouse

© Dr. Anab Whitehouse

Interrogative Imperative Institute

Brewer, Maine

04412

All rights are reserved. With the exception of material being used in compliance with the 'Fair Usage' clause of the Copyright Act, no portion of this publication may be reproduced in any form without the express written permission of the publisher. Furthermore, no part of this book may be stored in a retrieval system, nor transmitted in any form or by any means -- whether electronic, mechanical, photo-reproduction or otherwise -- without authorization from the publisher.

Published 2018

Printed in the United States

Published by: Bilquees Press

"Nothing is so fatal to the progress of the human mind as to suppose our views of science are ultimate; that there are no new mysteries in nature; that our triumphs are complete …"

Humphrey Davy

Table of Contents

Chapter 1 -- Constant Mysteries - page 7

Chapter 2 -- Antimatter Asymmetry - page 35

Chapter 3 -- All Tangled Up - page 69

Chapter 4 – Massive Problems - page 111

Chapter 5 -- Stringing Us Along - page 145

Chapter 6 -- Quantum Illusions - page 173

Chapter 7 -- Physical Conundrums – page 203

Chapter 8 -- Searching for Unity – page 235

Bibliography – page 253

Quantum Queries

6

| Quantum Queries |

Chapter 1: Constant Mysteries

If one were able to drill down to the deepest depths of physical reality, what would one find? In the 5th century BC, two Pre-Socratic thinkers – Leucippus and his student, Democritus -- maintained that one would discover nothing but indivisible atoms and empty space.

The answer to the question with which the previous paragraph began has changed a little bit over the last 2500 years. From the perspective of the modern concept of atoms, such material entities have been found not to be indivisible, and space might not be as empty as Leucippus and Democritus supposed but, rather – as some modern, theoretical heirs to the two Greek thinkers believe -- space might give expression to some sort of frothing, energetic dynamic.

However, Leucippus and Democritus weren't working with a Periodic Table consisting of – for now -- 118 elements. While their notion of an atom was something that was basic and fundamental (through which the diversity of material reality was constructed) nonetheless, such foundational atoms likely did not necessarily resemble or reflect the concept of an atom that exists today.

In fact, If Leucippus and Democritus had known about quarks and leptons, the two individuals might have felt that those particles were much more akin to what they had in mind when they spoke about 'atoms' than was the modern notion of atom. Moreover, if it turns out that quarks and leptons are not fundamental particles, then, Leucippus and Democritus might be inclined to favor the subunits from which quarks and leptons possibly arise as being more like the 'atoms' those two individuals envisioned some two and a half millennia ago.

Whatever the identity of the most basic entity of physical reality turns out to be – if, in fact, there is one entity … or any such basic entity at all -- should one consider the fundamental unit(s) of matter to be a function of particles or a function of forces?

| Quantum Queries |

Einstein's famous equation: $E=mc^2$ indicates that matter and energy are convertible one into the other, but that equation doesn't explain how energy and matter assume one modality of order rather than another, and, consequently, the basic unit of matter might not be, strictly speaking, either a particle or a force but something capable of assuming either the form of a particle or a force under different conditions.

The foregoing possibility resonates somewhat with a modern-day convention that refers to particles as "wavicles". In other words, the phenomena to which the term "particles" are given appear to exhibit both wave-like and particle-like properties under various circumstances, and, therefore, whether fundamental particles – whatever their names – constitute some kind of structured energy or some form of energized material form is uncertain.

Currently, physicists are seeking a unified theoretical model in which the four currently acknowledged forces (electromagnetic, weak, strong, and gravitational ... although Einstein would not have considered gravity to be a force but an expression of spatial geometry) constitute different manifestations of some underlying, fundamental force that can be induced (through, for example, the breaking of an initial symmetry) to give expression, under various physical conditions, to, at least, four different dimensions inherent in the potential of such a symmetry. On the other hand, there are boson particles that are described as being carriers of different kinds of forces – for example, photons are said to carry electromagnetic forces, gravitons (which no one has seen so far and, therefore, are theoretical) supposedly carry gravitational forces, gluons allegedly carry the strong force that binds quarks together, and the W^+, W^-, and Z bosons are involved with the weak force that plays a role in such phenomena as radioactivity.

How bosons 'carry' forces – or if they do -- is not really known. Conceivably, bosons might not actually carry forces as much as bosons constitute loci of manifestation through which forces emerge, or, maybe, bosons are merely structural components of various forces and, as such, bosons are not so much particles that carry force but are, instead, part of the way in which forces manifest themselves.

Alternatively, sometimes the notion of a force is described through the language of absorption and emission. In other words, when photons, for example, are absorbed or emitted by some interaction involving certain kinds of particles (e.g., electrons), then, this process of absorption or emission gives expression to the presence of – in this case -- electromagnetic forces that can be measured in various ways.

The mathematics of forces and particles constitute a description of a dynamic that can be hermeneutically parsed in terms of forces and/or particles, but the mathematical formulation is about the behavioral description of a given phenomenon over time -- or at a given instant of time -- and, consequently, such a mathematical formulation does not necessarily explain the ontological underpinnings that account (allegedly) for why a given phenomenon has the behavioral properties or dynamics that it does.

As such, the nature of many mathematical explanations tends to be limited. Those 'explanations' provide an account of how a given dynamic unfolds over time, but they do not necessarily account for what makes such a dynamic – characterized by a given set of properties -- possible in the first place.

Terms such as "force" or "particle" are the means through which concepts or ideas take on a linguistic form. The language of particles and forces are an attempt to make sense of the behavior/dynamic being described by a given system of mathematics.

Reliable calculations can be made through various systems of mathematics. Nonetheless, even though the foregoing sorts of calculations can illuminate, to varying degrees, the law-like properties of a phenomenon's dynamic, those calculations do not necessarily tell us what is actually going on with respect to a given phenomenon in any sense other than a behavioral sense.

In other words, mathematics cannot necessarily be used to tell one what a phenomenon is in a fundamental, ontological sense. Instead, mathematics tends to be used as a way of helping to describe what a given phenomenon does, or what values a phenomenon exhibits, under certain circumstances.

The phenomena being described by mathematics might only be shadow-like in nature. If this is the case, then, these sorts of silhouettes are merely surface forms that dance to the tune of a deeper reality, and while mathematics is capable of describing the nature of the dance, nevertheless, it fails to address questions that revolve about the issue of what makes such a dance possible or why the dance has the characteristics it does.

For example, there is a set of measurements that collectively are referred to as "constants" that are part of the quantitative description of various aspects of physical reality. Over the years, various experiments have been able to pin down the precise values -- more or less -- of such constants, but, to date, no one knows why those constants have the values they do ... and, therefore, although the constants can be used in the description of certain kinds of phenomena, those constants do not, yet, fit into any kind of explanatory system that accounts for why this universe is characterized by the presence of constants with the values that have been determined experimentally.

For instance, modern scientists know that the speed of light in a vacuum is a constant with a value of 299,792,458 meters per second. However, at the present time, there is no one within the world of science who can explain why light travels with that particular constant velocity under the specified conditions.

Of course, one could try to argue that there is no reason why the speed of light in a vacuum is 299,792,458. This is just the way things are.

There are several problems with the foregoing sort of perspective. First, at the present time, there is no available evidence capable of demonstrating that the foregoing position is correct, and, therefore, such a perspective is an entirely arbitrary one ... there is nothing scientific about it.

Secondly, to claim that the constants are just the way the ontological cookie crumbles, tends to run against the intuition or sense of existence that many, if not most, people seem to have concerning the nature of reality. One does not have to read any sort of theological or teleological meaning into the foregoing sentence since the intuitive sense to which allusions are being made has been at the heart of all kinds of attempts (including scientific ones) to understand the nature of reality and why phenomena have one set of characteristics rather than some other set of characteristics.

Perhaps, someone, someday, might prove there is no underlying reason why constants, among other realities, have the properties they do. Currently, however, writing constants off as being just the way things are seems a little premature ... even from a scientific perspective

Those who are seeking a scientific 'theory of everything' hope that the equations giving expression to such a theory will provide an account of how various physical constants – like the speed of light – derive their values from first principles. However, thus far, there is no theory of everything, and, therefore, the origin of the values displayed by various constants is hidden in the cloud of unknowing that permeates many aspects of modern, scientific understanding.

Physical constants come in several varieties. One edition is a function of quantities that involve concrete dimensions that are specified, while the other version of constants is referred to as being dimensionless.

The speed of light is one example of a dimensional physical constant because the value of that constant entails units of measurement that do not cancel out when expressed numerically. Another dimensional physical constant – the gravitational constant -- is used in helping to calculate the gravitational attraction that is exerted between two objects – namely, (as of 2014) 6.67191 × 10^{-11} $m^3kg^{-1}s^{-2}$, or, approximately, 6.67191x×10^{-11} Newtons $(m/kg)^2$ ... a value that was first approximated – remarkably well, as it turns out, -- by Henry Cavendish in 1798, 71 years after Newton passed away.

Dimensionless physical constants, on the other hand, involve units of measurement that cancel out during the process of calculations, leaving just a pure number. For example, the meters, seconds, and kilograms of the gravitational constant can be eliminated if one multiplies the gravitational constant by the mass of a proton squared, and, then, divides that numerator by the product of Planck's constant, h, and the speed of light, c.

Although the latter sort of calculation appears to be rooted in a somewhat arbitrary process, it does yield a number that provides a sense of how weak the force of gravitation actually is – if it is a force. More specifically, once the aforementioned calculation is carried out, one discovers that the relative magnitude of the coupling strength generated by the gravitational force is of the order of 10^{-38}.

The fine-structure constant is another dimensionless constant and is designated by: α. This constant gives expression to the relationship among the speed of light (c), Plank's constant (h), and the electron charge (e) -- all of which involve measurable dimensions of one kind or another. However, when combined together [usually according to some variation on, or related to: e/hc, the dimensions cancel out, leaving a value of about: 7.29735257×10^{-3}.

The fine-structure constant refers to the coupling strength of the electromagnetic interaction of elementary charged particles and brings together three of the central themes of modern science ... namely, quantum theory (h), the special theory of relativity (c), and electrodynamics (e). Arnold Sommerfeld first introduced this constant in 1916.

The reciprocal of the fine-structure constant – that is, α^{-1} – is the more usual (and more convenient) way of expressing this physical constant. The reciprocal of the fine-structure constant is 137.035999074 or approximately 1/137.

As long as the dynamic relationships among the components of the fine-structure constant remain the same, then the individual constants making up those components might assume different dimensional values under various circumstances, and, nonetheless, many aspects of electromagnetic interaction would not change – that is, such dynamics would appear to be observationally indistinguishable from what takes place in the universe with which we are currently familiar. However, if the fine-structure constant changed appreciably [meaning that the relationship of its components (e, h, and c) had shifted in some manner], then, there would be noticeable differences in the way in which electromagnetic dynamics manifested themselves under various conditions.

For instance, since the value of α – the fine-structure constant -- characterizes the strength of interacting particles, it has a dynamic role to play at the very heart of, among other things, the structural dynamics of atoms and molecules. If the value of α were not constant, then, the way atoms and molecules arise and interact with one another (including biologically) would be very different than what appears to be taking place around us today, and, as well, the stability of atomic and molecular interactions would be affected by changes in the value of α.

As indicated previously, Planck's constant constitutes one of the three components that form the relationship that gives expression to the fine-structure constant. The value of Planck's constant – designated h – is $6.62606957 \times 10^{-34}$ m²kg/s or approximately 6.626×10^{-34} Joule seconds.

| Quantum Queries |

In 1900 Max Planck came up with an idea born in desperation. On the one hand, he was trying to find a way around the problem posed by what was known as the Rayleigh-Jeans catastrophe (and later on – 1911 – it was dubbed the "ultraviolet catastrophe" by Paul Ehrenfest) and, on the other hand, Planck was trying to resolve the problem that arose when one considered the movement of an electron about the nucleus of an atom to be like that of a planet about a sun.

The ultraviolet catastrophe emerged when physicists made certain kinds of theoretical predictions. More specifically, when one calculated the amount of radiation that supposedly – at least according to classical physics -- would be emitted by an ideal black body at thermal equilibrium, one came up with an infinite result in the case of very short wavelengths, but experimental evidence indicated that nothing of the sort occurred.

Ideal black bodies are physical systems that are capable of absorbing all forms of radiation that engage it. When the black body re-radiates such absorbed energy, the amount of energy that is radiated in any given frequency range should be proportional to the number of resonance modes inherent in that frequency range, and, moreover, as the frequency range of the radiated energy approached the values of ultraviolet radiation (short wavelengths), the number of modes increased in proportion to the square of the frequency.

The theoretical predictions of classical physics indicated that the amount of emitted radiation should become infinite when an ideal black body emitted ultraviolet radiation. Experimental evidence contradicted the theoretical predictions, and Planck was trying to find a way to reconcile theoretical understanding with experimental data.

The problem that arose when one likened the movement of a electron to a planet around the sun concerned the ultimate fate of an electron if it moved in accordance with the idea of an orbiting object in a traditional or classical sense. The foregoing sort of orbiting movement involves acceleration, and this, in turn, means that the changing electrical field generated by an orbiting electron should be emitting photons as a function of such changes in the electrical field, and this, in turn, indicates that the electron should continuously lose energy to the point where the electron spirals into the nucleus.

Planck's condition of desperation in conjunction with the Rayleigh-Jeans catastrophe and the problem created by the alleged orbital motion of electrons around the nucleus of an atom led him to entertain a possibility that was anathema to the classical world view. Perhaps the energy in black bodies and 'orbiting' electrons could only take on certain quantized values.

By making the energy of a system proportional to the frequency of that energy's vibrational mode, and by making the behavior of electrons, atoms, and molecules a function of quantized units, Planck was able to come up with a formula that permitted him to perform calculations that were capable of accurately reflecting experimental data. Furthermore, by using his formula, Planck was able to determine the value of the quantized unit (i.e., $6.62606957 \times 10^{-34}$ m^{2}kg/s or approximately 6.626×10^{-34} Joule seconds) that gave expression to the basic unit of manifested energy or action for electromagnetic energy.

As the frequency of a wavelength increased, then, so did the number of packets of energy associated with such frequencies. Increases and decreases in wavelength were a function of the multiple emission and absorption of such units of energy or action.

If there were quantized limits on how black bodies radiated energy at different frequencies, and if there were quantized limits on the relationship between electrons and the nucleus of an atom, then, several outstanding problems in physics (described earlier) could be resolved. However, this manner of resolving things depended on the basic unit of action or energy having a constant, discrete, minimal value -- h -- Planck's constant.

Initially, Planck considered his formula to be the stuff of desperation and merely a way to avoid some problems. However, over the next 75 years -- beginning with Einstein's 1905 explanation of the photoelectric effect that both presupposed, as well as constituted evidential support for, Planck's discovery of desperation -- the idea of quantum phenomena that were rooted in the notion of a fundamental unit of action or energy took center stage in helping to provide reliable explanations for many physical phenomena.

The existence of Planck's constant permitted many kinds of problems to be solved and many kinds of calculations to be made. Nonetheless, no one knew why the basic unit of action or energy should have the value of $6.62606957 \times 10^{-34}$ m²kg/s or approximately 6.626×10^{-34} Joule seconds.

What, if anything, set that minimal, constant value? How did systems "know" how to emit or absorb quanta of precisely that value, and what were the underlying dynamics, if any, of such a process?

Besides Planck's constant, another one of the three components that give expression to the fine-structure constant discussed previously is 'electron charge'. According to an arbitrary convention introduced by Benjamin Franklin, the charge of an electron is said to be "negative", and that charge has been determined experimentally to be 'roughly' (just kidding): $1.60217657 \times 10^{-19}$ coulombs ... a dimensional constant.

Supposedly, electrons give expression to an entity that is devoid of any internal structure. At the same time, electrons can be described through four quantum numbers that collectively characterize an electron's behavior within an atom.

Those four numbers are: (1) the principle quantum number, n, that identifies the electron's energy level and, therefore, serves as an index for the electrons most likely distance from the atom's nucleus; (2) the orbital angular momentum number, l, that determines the shape of the orbital state within which an electron resides, and, therefore, constitutes an index for the distribution of angular momentum within the orbital; (3) the magnetic quantum number, ml, that designates the number of orbitals, along with their orientation, within a given subshell and is dependent on the value of the orbital angular momentum; and, (4) the electron spin quantum number, ms, that can take on the values of either +1/2 or -1/2 and plays a role in determining whether, or not, an atom will have the ability to generate a magnetic field.

How a particle without any internal structure is capable of giving expression to a phenomenon that can be described by changing quantum numbers and, as well, by a constant electric charge is not known. Perhaps, the quantum numbers are emergent properties arising through the interaction of an electron with the atom that houses it, but this still tends to imply that there are structural features inherent in an electron that can be induced to manifest themselves in different ways under changing conditions of atomic and molecular dynamics.

What is known about the electron concerns the parameters of its behavioral dynamics under different circumstances rather than its structural character (if any). Nevertheless, we really don't know what makes such behavioral properties possible, and we don't understand what permits different electrons to exhibit the same, precise, constant charge value under varying conditions.

However, claiming that electrons are point-masses (i.e., entities with no internal structure), might be something of a simplification that obscures what could turn out to be a more complex reality. For example, according to some modern interpretations and applications of quantum theory, space is not empty but constitutes a bubbling, churning dynamic of virtual electron-positron pairs -- along with assorted virtual photons -- that blink in and out of existence within very small intervals of both space and time.

The foregoing sorts of virtual particles are believed to swirl about any given electron. As a result, there is a process of polarization that allegedly takes place in and around electrons as virtual electrons are repelled by non-virtual electrons, while virtual positrons arising from the dynamic that constitutes space are attracted to -- or by ... (or both) -- non-virtual electrons, and, in the process, a surplus of virtual positrons are believed to gather about non-virtual electrons.

Quantum theory claims that the foregoing polarization process creates a cloud of virtual particles around the non-virtual electrons. In the process, a portion of the non-virtual electron's charge is modulated or dampened by the swirling cloud of virtual particles that surround the non-virtual electron.

Considered from the foregoing perspective, electrons are not point-masses. Instead, they are structured complexes consisting of a non-virtual electron and its virtual companions.

Treating electrons as structured entities rather than point-masses has some benefits. For example, if one tries to calculate the charge of an electron in the absence of its virtual cloud, one arrives at a value that is infinite, whereas actual measurements involving non-virtual electrons produce a finite value ... the aforementioned **$1.60217657 \times 10^{-19}$ coulombs.**

As nicely as the foregoing structured perspective on electrons works out as far as mathematical calculations are concerned, there are some problems that seem to swirl about such a theoretical position like so many virtual particles swirling about a non-virtual electron. For example, the structured notion of an electron is based, in part, on the assumption that space is a dynamic involving the generation of virtual particles, but there is no concrete evidence to indicate space constitutes a process that generates virtual particles or which indicates that space is responsible for such a process.

The alleged capacity of space to give expression to the generation of virtual particles might just be a theoretical artifact of the mathematics of quantum physics that has led some physicists to believe there are ontological realities occupying the Heisenbergian uncertainties associated with quantum probabilities instead of holding that the mathematics of uncertainties and probabilities associated with quantum phenomena might just constitute some of the epistemological limits of quantum methodology and, as well, might actually obscure the deep character of reality as being something other – as Einstein believed – than a random, probabilistic phenomenon. In other words, quantum theory might be mistaking its own probabilistic methodology for the nature of ontology, and, in the process, assuming that the quantum nature of space harbors the potential for generating an indefinite, if not infinite, supply of virtual particles.

Is space a quantum phenomenon? We don't know!

Whatever space turns out to be, should one automatically suppose that it has the capacity to generate virtual particles? Not necessarily, and, at the very least (and as was previously noted), currently, there is no evidence to indicate that space is in the business of either generating virtual particles or that the fabric of space constitutes a quantum dynamic.

Conceivably, the generation of virtual particles takes place independently of space. Perhaps, there is some sort of dynamic that is contained within space or that manifests itself through space that is the actual source (rather than space) of the generation of virtual particles ... to whatever extent this actually takes place.

Even if the generation of virtual particles were associated, in some fashion, with the structural, dynamic character of space, one wonders why virtual positrons would <u>cloud about</u> non-virtual electrons instead of engaging in the process of <u>mutual annihilation</u>. Perhaps, the virtual particles are not in existence long enough to annihilate their non-virtual counterparts, but, if this is the case, then, maybe, the virtual particles also are not in existence for a sufficiently long enough period of time to be able to enter into the sort of interactions that would create polarized conditions in and around non-virtual electrons.

Where do the forces of attraction leave off and the forces of annihilation begin? Or, is it possible for any kind of attraction dynamic to exist between antiparticles without, simultaneously, a tendency toward annihilation being present as well?

If the interactions between electron and positron are anything like the interaction between protons and antiprotons, then the process of annihilation is not immediate. Instead, there seems to be a little dance of death that transpires before an inevitable fate catches up with interacting protons and antiprotons.

More specifically, when the respective fields of charge from a proton and an antiproton begin to resonate with one another due to their proximity, they revolve about each other. In the process, gamma rays are emitted indicating that a loss of energy is taking place and, as a result, the two antiparticles are approaching one another until, eventually, mutual annihilation occurs.

The delay in the process of annihilation in relation to proton and antiproton interaction might be a reflection of the relative complexity of the internal arrangement of quarks in protons and antiprotons. Perhaps, because of that internal structure, it takes time (brief though this might be) for forces and charges to align themselves in a way that culminates with annihilation.

If electrons and positrons have some sort of internal structure, then as is the case with protons and antiprotons, there might be a brief dance of death involved in the interaction of such antiparticles as there is in the case of protons and antiprotons. If this is the case, then, one wonders, if there will be some indication (as the presence of gamma radiation indicated in the case of interacting protons and antiprotons) that there is a loss of energy taking place between the interacting electron and positron prior to their eventual annihilation.

Maybe, the dampening effect concerning the charge of an electron that has been experimentally observed and that has been attributed to a cloud of antiparticles surrounding an electron due to a process of polarization of space that supposedly occurs is actually a function of some sort of sub-structural dynamics taking place and that interferes, to some degree, with the amount of charge that can be measured under different circumstances.

Moreover, if virtual pairs are constantly being created and destroyed due to the capacity of space to generate paired quantum processes, then how does the virtual particle/non-virtual electron polarization dynamic affect the annihilation process? After all, if virtual positrons are busy being polarized and, in the process, swirl about non-virtual electrons without destroying the latter, then, how do such polarized virtual particles meet up with appropriate virtual anti-particle partners in order to find their way back to oblivion?

Virtual particles might be created in pairs. However, quantum theory seems to be proposing the possibility that there is an asymmetry introduced into reality through the polarization process that takes place between non-virtual electrons and virtual positrons, and, therefore, this raises questions about the nature of the dynamic through which virtual particles blink back out of manifested existence.

One might also wish to entertain a variation on the: 'How many angels fit on the head of a pin' question in conjunction with the foregoing considerations. More specifically, how many virtual positrons can fit into the cloud surrounding a non-virtual electron?

Are there inherent limits to the structural character of such a cloud? What maintains the cloud if positrons are being 'pulled' out of the cloud in order to return to the state of annihilated being with their virtual antiparticles?

Are virtual particles precisely the same as their non-virtual counterparts? For example, do virtual particles have the potential to be characterized by different quantum numbers, or are those quantum numbers a functional artifact of the way an electron interacts in the presence of an atom, and, if the latter is the case, then, how does the presence of an atom tap into, and induce, the properties associated with an electron's quantum numbers to become manifest?

The alleged capacity of virtual positrons to polarize the space in the vicinity of non-virtual electrons without annihilating the electrons around which they gather raises the possibility that, maybe, there are certain differences between virtual particles and non-virtual particles. If so, then, perhaps, as a result of such differences, the constant that characterizes, say, the charge of non-virtual particles is not present in virtual particles … or not present to the same degree.

On the other hand, if virtual particles are the same as their non-virtual counterparts, then, one wonders how the constancy of charge arises out of the allegedly random process to which the generation of virtual processes supposedly gives expression. Furthermore, one also wonders why the presence of all these clouds of virtual particles doesn't interfere with the dynamics and stability of an atom … or, maybe, it does, and this – along with the weak force -- is part of what underlies, say, the phenomenon of, radioactivity.

No one has ever seen a virtual particle. Rather, the notion of virtual particles is a way of trying to make sense of various observed phenomena.

The mathematics of quantum mechanics describes events in which different kinds of quantitative change take place in the properties of interacting particles. Some of those changes are attributed to the exchange of virtual particles or, alternatively, are attributed to the emission and absorption of virtual particles.

| Quantum Queries |

However, the actual nature of the exchange, emission, or absorption process is never seen. Perhaps, the processes being described reflect the existence of virtual particles, and, then again, maybe, those processes give expression to some other kind of dynamic that does not involve virtual particles but for which the concept of 'virtual particles' serves as a hermeneutical surrogate.

One doesn't have to reject quantum mechanics in order to question the ontological status of virtual particles. Quantum mechanics has proven itself to be a very accurate mathematical method through which to describe, among other things, the behavioral dynamics of interacting particles, but what is being described in conjunction with such interactions might, or might not, be a function of the presence of virtual particles.

Non-virtual particles surrounded by clouds of virtual particles make mathematical sense. Non-virtual particles that are not surrounded by clouds of virtual particles do not make mathematical sense because non-virtual particles devoid of clouds lead to the calculation of embarrassing infinities.

The experimental measurements associated with non-virtual particles are all finite. Non-virtual particles absent their virtual clouds seem to entail infinities of one kind or another (e.g., charge and mass).

Following Paul Dirac's 1928 discovery of an equation that brought together the theories of special relativity and quantum mechanics, Dirac used his equation to determine that electron's must have a magnetic moment. Dirac calculated a theoretical value for that magnetic property, and his result reflected the experimentally determined value that was available at that time

Several decades later, a new experimental determination of the value of the electron's magnetic moment was made, and it diverged from Dirac's original theoretical calculated value. The difference was fairly small – about 1% -- but the difference, small though it might be, called for an explanation.

Eventually, the difference between experimental and theoretical values involving the electron's magnetic moment was attributed to the dynamics of the virtual particles that allegedly swirled about the electron. Moreover, by using the mathematical techniques of renormalization theory, many of the infinities associated with such dynamics could be cancelled out and, as a result, one was left with only the value of the experimentally determined results for the magnetic moment.

Many people are familiar with how various quantities and values can be cancelled out during the process of making mathematical calculations when trying to solve a given problem. Such possibilities are first encountered in high school math and physics.

On the other hand, however, understanding how infinities cancel out ontologically – and not just mathematically -- is a little more challenging. For instance, presumably an infinite amount of time is not required for infinite quantities to cancel one another, but, if such cancellations are instantaneous (or relatively so), then what is the nature of the dynamics that causes such relatively instantaneous, ontological cancellations to occur?

The mathematics of renormalization might permit one to obtain quantitative answers that are not embarrassing. Nonetheless, those techniques for arriving at sensible mathematical answers shed little light on the nature of the ontological dynamics that, supposedly, are being described through those techniques.

If one doesn't like using the idea of virtual clouds to give non-virtual particles the sort of complexity of structure that is needed to eliminate the problem of infinities, there is another possibility to consider. One could choose some arbitrary size – for instance, 10^{-18} centimeters – and claim that electrons possesses a very small but non-zero radius and, therefore, they have a structure of some kind at that arbitrarily selected level.

If, some day, scientists were able to experimentally probe the realm of 10^{-18} centimeters and still found no indication that electrons had a structure, then, one could just take the level at which structure in an electron had been hypothesized to begin to be visible and alter that hypothesized value from 10^{-18} centimeters, to, say 10^{-21} centimeters ... or, maybe, even 10^{-23} centimeters. After all, 10^{-21} or 10^{-23} centimeters is a long, long, long way from the size of being zero, and as long as the size is not zero, then, for example, any particle with mass or charge will not suddenly appear with an infinite density or an infinite charge.

Finally, whatever the realities of virtual particles might be with respect to whether, or not, they exist, or in relation to the manner in which they interact with, say, electrons, and, in the process, lend a structural dimension to electrons, there is still, at least, one further problem. More specifically, even if virtual particles cloud around, say, electrons, and, thereby, help eliminate the problem of infinities involving mass and charge, nonetheless, how such a clouding process will give rise to the other properties that involve the different quantum number values (see page 33) that are associated with, in this case, electrons is still unclear.

Considered from, yet, another perspective, one could treat the electron as a point-mass (i.e., possessing zero size) without necessarily having to encounter the infamous infinities of earlier calculations. For instance, if the electron were merely the phenomenal point through which a deeper reality manifested itself, then, such a point does not so much contain a mass, charge, spin, and so on, as much as that locus of manifestation gives physical expression to the dynamics of another dimension or set of dimensions beyond the space in which the phenomenal electron is manifested ... just as a holographic image doesn't actually contain any of the material properties that are being projected onto a given region of space through the way coherent light gives life to stored information that is independent of that holographic image.

Of course, even if one were to accept something along the lines of the foregoing notion, one is still left without an explanation of why the charge of an electron is constant. In addition, one is still left without an account of how the constancy of the electron charge is generated and maintained.

Scientists speak of the aforementioned dimensional and dimensionless values as being constants. However, whether, or not, those values actually are constant has not been established in any final sense.

Certainly, at the present time, there is no hard evidence available indicating that the constants are not really constant. Nonetheless, the current lack of evidence in this regard has only encouraged some researchers to explore that possibility.

While individuals like Einstein believed the properties of constants were set by the nature of interactions, other theorists (e.g., Paul Dirac) have entertained the possibility that the values of the so-called constants might, for example, diminish with the passage of time – that is, become smaller as the Universe grows older. However, irrespective of whether, on the one hand, the nature of physical interactions fix certain values as constants (or, could it be that constants are the reason why physical interactions give expression to universal laws?), or whether, on the other hand, 'constants' are subject to some degree of change, there seem to be as many problems surrounding the questions of how and why certain values remain constant as there would be problems in conjunction with the issue of why seemingly fundamental properties -- such as the speed of light, the gravitational constant, and the fine-structure constant -- might change.

If certain values -- such as the speed of light, the coupling strength of gravitation, the fine-structure relationship, α, as well as the nature of about 17 other physical values -- are, have been, and will continue to be constant, then physics is faced with a huge challenge. Indeed, if constants truly are constant, then, physicists will have to come up with a theory that is capable of showing how the values of all the foregoing constants can be derived from first principles ... a set of derivations that seems to reside somewhere far beyond the horizons of current understanding.

Moreover, if physicists cannot come up with a theory that entails the capacity to derive the precise value of the foregoing constants from first principles, then, there will be fundamental facets of reality that physics does not encompass and, as a result, physicists will not be able to explain why the universe is the way it is. Under such circumstances, constants could be used to help <u>describe</u> the universe, but the use of those constants in one's description of various phenomena would be very limited in their capacity to help one to <u>explain</u> why the universe is the way it is.

If we don't know why constants have the precise values they do, then, we really don't know how the universe came to be the way it is or why the universe exhibits the constants it does. Having to hand-feed constants into calculations is an indication of our collective ignorance ... an indication of just how little the curtain of mystery has been drawn back with respect to the nature of reality.

On the other hand, if particular values that appear to be fundamental to the way the universe operates are not constant, then, physicists will be faced with an equally daunting challenge. They will have to account for how and why certain degrees of freedom seem to be built into the fabric of the universe in such a way that a number of seemingly fundamental physical values have been permitted to remain sufficiently constant for considerable periods of time (say, for 14 billion years) so that the manner in which physical dynamics take place does not seem to have changed appreciably.

Alternatively, if so-called constants are able to change in more than very minor ways, then, one cannot necessarily suppose that the laws of physics are universal in character since those laws depend, in critical ways, on an array of constants retaining their values. As a result, if constants do change, over time, beyond certain minimal degrees of freedom, then physicists might have a problem explaining how we got here (today) from there (some time in the past) since one can no longer assume that the physical laws that governed some facet of the past are the same physical laws that appear to be in evidence today.

Is there any evidence that the constants of nature – or, at least, some of them – change? Possibly, but the data (and related interpretations) are very iffy in several ways, and even if such data were considered to be definitive, nonetheless, the parameters within which the change in some constants might take place are so constrained that there would not necessarily be any appreciable difference in the physical properties of the world.

For example, consider the ratio of uranium 235 to uranium 238. In naturally occurring ores, uranium 235 tends to constitute only a small proportion of the uranium in any given ore deposit.

Uranium 238 is not capable of generating a self-sustaining nuclear chain reaction. Uranium 235, on the other hand, is capable of generating such reactions when the concentration of that isotope relative to uranium 238 reaches about 20 per cent (weapons grade uranium requires about a 90 per cent concentration of uranium 235 relative to uranium 238).

Generally speaking, the natural occurring ratio of uranium 235 to uranium 238 that has been determined from a variety of both earthly and cosmic (e.g., lunar samples and meteorites) sources is about 0.720 per cent. In 1972, however, a French scientist, Dr. Henri Bouzigues, analyzed some uranium hexafluoride gas samples and obtained a measurement that differed with the well-established uranium 235 to uranium 238 ratio noted earlier – namely, 0.720.

The uranium ratio result obtained by Dr. Bouzigues was 0.717 per cent. 0.003 per cent of the usual amount of uranium 235 appeared to be missing.

As indicated previously, the uranium ratio value of 0.720 for uranium 235 and uranium 238 is well established and has been measured in many different kinds of contexts both earthly and unearthly (e.g., lunar samples and meteorites). What was going on?

The uranium samples being tested by Dr. Bouzigues were from uranium deposits located near the Oklo River in the West African nation of Gabon. After entertaining many possibilities in an effort to account for the 0.003 per cent discrepancy ratio, investigators discovered that all the shipments of uranium that had been coming from the Oklo region since 1970 showed the same slight depletion in the uranium 235/238 ratio readings that Dr. Bouzigues had discovered.

Upon further analysis, researchers found that all the Oklo uranium samples collected between 1970 and 1972 displayed a familiar signature. More specifically, those samples revealed the presence of more than 30 by-product elements that are characteristic of nuclear fission having taken place in relation to the Oklo uranium deposits.

When the foregoing research was completed, 14 sites had been found at the Oklo River uranium complex that exhibited signs of having served as the operating location for naturally occurring nuclear reactors approximately two billion years ago. In other words, the missing uranium 235 had been used up helping to sustain a chain of nuclear reactions that occurred at some point in the distant past.

The half-lives of uranium 235 and uranium 238 are different. Uranium 235 has a half-life of about 700 million years, while the half-life of uranium 238 is approximately 4.5 billion years.

When the Earth first came into existence some 4.5 billion years ago, the ratio of uranium 235 to uranium 238 was roughly 17 per cent ... many times higher than is the case today. However, over time, the amount of uranium 235 relative to uranium 238 began to lessen as a function of the quicker half-life of the former isotope relative to uranium 238.

Due to the differing half-lives of the two uranium isotopes, the ratio of uranium 235 to uranium 238 dropped about 14 per cent during the next 2 and one-half billion years. When the ratio reached 3 per cent, uranium 235 became capable of initiating -- with a little bit of help from surrounding conditions -- a sustained chain of nuclear reaction.

Among the conditions needed to help ignite a sustained nuclear reaction were some geological ones. For example, granite strata in the Oklo region happened to be tilted at an angle (approximately 45 degrees) and, as a result, under the right circumstances, ground/rain water could accumulate and serve as a solvent for deposits of uranium ore, and eventually, this would lead to the formation of uranium oxide.

If the volume of the accumulated water remained the same, this would provide uranium oxide with an opportunity to increase in concentration. Under certain conditions (to be discussed shortly), the water also helped to slow down neutrons being released by the decay of uranium nuclei so that the neutrons could be absorbed by uranium 235 and not just by uranium 238 or other kinds of neutron absorption traps.

There were several other conditions that were needed to help sustain a nuclear reaction once it started. For instance, the concentration of uranium oxide in the accumulated water needed to be about 10 per cent, and, in addition, the seam of uranium deposits that were present had to be at least a half a meter in thickness in order to prevent the neutrons being released during the early stages of nuclear reaction from escaping.

As the nuclear reaction sped up, more energy -- in the form of fast moving neutrons -- was released and the temperature of the surrounding water would rise ... eventually producing steam. The steam had the effect of slowing down neutrons, and the

| Quantum Queries |

slowing down of neutrons led to a fall in the temperature of the water, and the decrease in water temperature brought about the condensation of the steam back into a liquid form and, in the process, reduced the number of neutrons being absorbed.

The rise and fall of water temperature worked like graphite rods with respect to the movement of neutrons in a man-made nuclear reactor. When other conditions were met -- such as the right concentration of uranium oxide, and the thickness of uranium deposits -- a naturally occurring, sustained, nuclear chain-reaction could take place ... and did so on a number of occasions in the Oklo region several billion years ago.

What does any of the foregoing have to do with the issue of constants? Possibly, it might have a lot to do with the issue of constants ... or, at least, some of them.

Alexander Shlyakhter, a Russian physicist, noted something in the Oklo uranium research that he considered to be intriguing. One of the nuclear reactions that took place several billion years ago was of a very unique nature.

The reaction that interested Shlyakhter involved the absorption of a neutron by samarium-149 nucleus. This led to the formation of the samarium-150 isotope plus the release of a photon.

However, the absorption of the neutron was functionally related to the presence of a form of resonance in samarium-149 that was sensitive to a very narrow range of energies. The specific character of the necessary resonance was dependent on the existence of a confluence of three forces: (1) electromagnetism, (2) the weak nuclear force, and (3) the strong nuclear force.

The probability that a neutron will be absorbed by samarium-149 varies with temperature as the resonance energy value is altered. If no shift were observed in the samples being analyzed, then, the resonance value of two billion years ago would be the same as the one that is observed currently, and, this would indicate that there has been stability over time in the value of the constants that make the foregoing resonance possible.

The problem with trying to compare possible differences in past and present samarium-149 resonance values is that due to the complexity of the dynamics involved in the interaction between a captured neutron and the nucleus of samarium-149, determining precisely how the aforementioned three forces are differentially contributing to the setting of the resonance value in samarium-149 is not presently possible. All that can be done is to establish parameters within which one or another of the fundamental forces might be changing the character of their contribution to the setting of the resonance value in samarium-149.

In calculating his results, Shlyakhter made the assumption that the three forces contributed to the setting of the resonance value in proportion to their strengths. He also assumed that the temperature in his naturally occurring reactor was 300 degrees, or so, Centigrade.

Given the foregoing assumptions, Shlyakhter concluded that there could have been a difference between the samarium-149 resonance value of today and the samarium-149 resonance value that existed two billion years ago. However, he believed that if there were such a difference, the variance could not have exceeded 20 milli electron volts relative to the current resonance value for samarium-149, and this translates into a possible change in the resonance value for samarium-149 of less than one part in five billion over several billion years.

Of course, if the temperature of the naturally occurring, two billion old, Oklo nuclear reactors were other than the 300 degrees Centigrade assumed by Shlyakhter, then, to some extent, the foregoing calculations would be affected. Most of the individuals who have researched the foregoing issue tend to agree that the temperature of the Oklo reactors ran somewhere between 200 and 400 degrees Centigrade, and, therefore, there

| Quantum Queries |

is a range of corrections that might have to be made with respect to Shlyakhter's original calculations.

In addition, if the three fundamental forces (electromagnetic, weak, and strong) did not contribute to the setting of the samarium-149 resonance value in proportion to the respective strengths of those forces -- as Shlyakhter assumed -- then his calculations also would be affected accordingly. While there is an indefinite number of possible combinations that are conceivable involving differing, relative contributions from the three fundamental forces (electromagnetism, the weak force, and the strong force) with respect to the setting of the samarium-149 resonance value, nevertheless, whatever values those combinatorics might yield, then in most, if not all of such cases, those values still would be relatively close to the change parameters being set by Shlyakhter in conjunction with the possible difference in values for the resonance energies associated with samarium-149 between 2 billion years ago and the present.

None of the foregoing possibilities indicate that any of the constants changed. Rather, the calculations indicate that if such changes did occur, then, they would have to fall within certain parameters if one is to be able to reconcile what took place several billion years ago with what is observed happening today with respect to the resonance value of samarium-149 that makes the formation of samarium-150 possible.

The Shlyakhter calculations, however, do give rise to a question that cannot be answered – at least at the present time -- even if one acknowledges that such parameters accurately describe the putative changes that might have transpired over the last several billion years with respect to the resonance energy value for samarium-149. More specifically, the unanswered question is this: What forces or modulating factors could have maintained -- for billions of years -- the variation in so-called constants (electromagnetism, the strong force, and/or the weak force) within parameters that, according to Shlyakhter's calculations, are quite small?

If constants really are constant, and do not change, then, one has the mystery of constancy. Under such circumstances, one is faced with the problem of determining the how and the why (if any) of that constancy ... a problem that modern physics is currently unable to resolve.

On the other hand, if 'constants' are not so constant but vary to some degree and do so within limited parameters, then one is confronted with a different kind of problem. Namely, what constrains such variance or degrees of freedom within the indicated parameters?

The latter issue is a problem about which modern physics is uncertain as to whether, or not, it is even a real conundrum. However, if that problem turns out to be real, then, at the present time, modern physics does not know what the solution to the problem of 'near-constancy' looks like.

Chapter 2: Antimatter Asymmetry

According to many proponents of the Big Bang theory of the universe, there must have been a slight asymmetry between the number of particles and the number of antiparticles that existed in the universe some 10^{-40} seconds – or there about -- following the alleged event that propelled the universe on a journey that supposedly has been taking place over the last 14 billion-plus years (more on this in Volume III). The size of the foregoing asymmetry has been calculated by some to consist of one extra particle of matter for every ten billion parings of matter and antimatter.

The abovementioned calculation is hypothetical since there is no available evidence to indicate that such an asymmetry actually existed in the early universe. Rather, the calculation gives expression to a figure that – if it were correct -- might be able to make sense of the fact that the present universe seems to be largely devoid of any signs of antimatter even though the early universe of 10^{-40} seconds is believed to have contained equal (or nearly so) proportions of both matter and antimatter.

Apparently, for each ten billion pairings of matter and antimatter particles that existed at 10^{-40} seconds and which annihilated one another, an additional particle of matter was hypothesized to have survived the cosmic carnage. Indeed, the universe we see today is supposedly made up of all of those extra, unpaired matter particles that were left over after all the other particle-antiparticle pairings met with oblivion during their dance of annihilation.

However, if events in the early universe (as far as the asymmetry of matter over antimatter is concerned) didn't transpire in accordance with the manner in which they were hypothesized to have taken place (as outlined above), then, there are a few problems. For example, why -- for the most part -- is only matter found in the universe, or, considered from a slightly different perspective, what happened to all the antimatter that allegedly existed in the time just before, during, and/or just after, the Big Bang occurred?

Even if one were to accept the hypothetical calculation noted several paragraphs ago, one is still left with some problems. More specifically, why was there any asymmetry at all in the pre- or post-10^{-40} second universe with respect to the relationship between matter and antimatter, and why did the asymmetry have just the properties needed (that is, an extra particle of matter for every ten billion particle-antiparticle parings) to give expression to the world we see today?

An obvious response to the last question is that if things just prior to, or just following, the Big Bang were not as some scientists have hypothesized them to be, then, the universe we witness today – the one that appears to be largely devoid of antimatter – doesn't seem to make a lot of sense. Indeed, unless one can provide some concrete evidence to support the foregoing hypothesis concerning extra matter relative to antimatter, then, such an approach to things is really little more than an exercise in assuming one's conclusions.

Of course, there are a lot of questions and mysteries surrounding the state of the universe at 10^{-40} seconds because that time apparently encompasses conditions that are beyond the understanding of current physics. One dimension of those conditions could involve the reason(s) why there might have been an asymmetry between matter and antimatter, and another mysterious dimension of the physical conditions prevailing at 10^{-40} seconds might touch upon the question of how matter and antimatter could have been brought together in such close confinement within the singularity that is believed to have preceded the Big Bang.

The aforementioned issue of close confinement at the time of (or just before) the Big Bang might be especially pressing if there really were ten billion pairings of matter and anti-matter for every particle of matter that survived the dance of annihilation that either helped fuel, or took place in conjunction with, the Big Bang. After all, the current universe consists of quite a lot of matter, and, consequently, if the aforementioned calculations concerning the asymmetry of matter over antimatter are correct, then, as far as numbers of particles are concerned, the singularity that many scientists believe preceded the Big Bang could have included an additional 20,000,000,000 particles for each and every particle that makes up the present universe ... all compressed or contained within the relatively small confines of the original singularity that allegedly preceded the birth of the present universe.

Did all of the foregoing particles exist prior to the Big Bang? Present understanding does not permit us to give a reliable answer to that question, but if all those particles and antiparticles did exist prior to the Big Bang, then one wonders how so many particle-antiparticle pairings coexisted within the relatively confined space of a singularity of some sort and one also wonders how long such conditions of relatively close confinement lasted.

Did only some of the foregoing particles exist prior to the Big Bang (assuming, of course, that there was a Big Bang) and, if so, how many of each kind of particle and antiparticle were there? Current scientific understanding doesn't permit a definitive answer to be given with respect to either of the foregoing two questions.

Were the foregoing particles created through the energy released by the Big Bang? No one really knows ... although some number of such particles might have been created via the alleged high-energy dynamics of a Big Bang.

Was the energy released by the Big Bang sufficiently great to be able to create 10,000,000,000 particle-antiparticle pairings for each particle that survived the Big Bang? At the present time, science cannot answer the foregoing question.

Was there a transition from energy to particle-antiparticle pairings during the Big Bang, and, if so, why would there have been an asymmetry created with respect to the amounts of matter and antimatter that were created? Up to the present time, science has not been able to provide reliable answers to either of the foregoing questions.

Assuming that at some point prior to, or during, the Big Bang, 10 billion pairings of particle-antiparticle (for each additional particle that survived to form the present universe) annihilated one another, when did that event or series of events take place, and what happened to the energy released by those annihilations? Currently, no one knows.

One doesn't have to visit the LHC (Large Hadron Collider) at CERN, in Switzerland (scheduled to go back on line in early 2015), in order to encounter antiparticles. For example positrons are generated during PET scans by a positron emitting radionuclide or tracer that is introduced into a person's body after the tracer has been attached to a biologically active molecule such as fluorodeoxyglucose.

Eventually, the positrons produced in the foregoing manner bump into electrons within a person's body. Such collisions annihilate both the positron and the electron and, in the process, generate gamma rays that can be used to construct a three-dimensional image of tracer concentration within a person's body.

Antiparticles also arise naturally when, under appropriate conditions, a proton transitions into a neutron, releasing energy during the transition. A portion of that released energy manifests itself as a positron.

In addition, positrons naturally arise in the heart of the sun where they collide with electrons that have been stripped from hydrogen atoms due to the enormous temperatures existing at the center of the sun (believed to be more than 10 million degrees). The annihilation of the two interacting particles releases gamma rays that travel away from the center of the sun at the speed of light but repeatedly lose energy through a series of encounters with electrons on the journey of the gamma rays to the surface of the sun (a journey that is considered to last a hundred thousand years) before radiating as sunlight that reaches Earth in about eight minutes and 20 seconds.

Finally, positrons also arise in conjunction with the high-energy, and still mysterious, cosmic rays that enter the Earth's atmosphere and collide with Earth-bound atoms of one kind or another. The resulting dynamic produces a cascade of, among other things, electrons and positrons.

Obviously, given the foregoing considerations, there are plenty of ways to generate antiparticles through various kinds of dynamics (e.g., high-energy physics experiments at CERN, PET scans, the interior of the sun, cosmic rays, protons transitioning into neutrons). However, antiparticles do not seem to exist on their own independently of the foregoing sorts of dynamics.

How did we come to live in a universe in which antiparticles arise as natural by-products of interactions involving non-antimatter particles, rather than come to live in a universe in which non-antimatter particles arise as natural by-products of antimatter particle interactions? How did we come to live in a universe in which matter appears to totally predominate, while antimatter seems to have an important, but subsidiary role to play that is constrained by, and largely dependent on, the dynamics of matter?

The perspective that prevails among many modern scientists is that whatever happened in relation to antimatter took place very soon after the Big Bang got underway. More specifically, within the first one millionth of a second, antimatter already was supposedly being led down the path to relative non-existence with respect to matter as well working its way toward assuming its subsidiary role in relation to matter.

The foregoing shift in status involving matter and anti-matter allegedly occurred sometime between 10^{-40} seconds and 10^{-12}

seconds following the Big Bang. So, what could explain such a transition?

At this point in time, science is limited to theoretical possibilities concerning how such a shift away from roughly equal numbers of particles and antiparticles might have taken place. In other words, currently, scientists do not possess the capabilities necessary to reproduce the conditions that might have existed between 10^{-40} seconds and 10^{-12} seconds following the Big Bang, and, consequently, scientists have to speculate about – rather than directly or indirectly observe -- what might have transpired during the foregoing very brief interval.

Although Dmitry Skobeltzyn, in 1923, was the first individual to observe experimental evidence revealing the existence of antimatter, he did not understand the significance of what he was witnessing in his lab, and, as a result, Paul Dirac was the first individual to propose the idea that something like antimatter might exist. His proposal arose out of the equation -- $i\gamma \cdot \partial \psi = m\psi$ -- he had discovered in 1928 that was consistent with the principles of both quantum mechanics and the theory of relativity.

In constructing his equation, Dirac resorted to using the mathematics of matrices to try to get around a difficulty that appeared to be implicit in Einstein's theory of relativity. In short, both positive and negative quantities seemed to be involved in solving various kinds of problems concerning the energy associated with the dynamics of bodies at rest and in motion.

More specifically, according to Einstein's theory of relativity, a body in motion has an amount of energy (E) that is proportional to the area of a square with a side that forms the hypotenuse of a right angle triangle with a base that is proportional to the amount of energy in such a body when at rest (mc^2), and according to Einstein's theory of relativity, the energy of a body in motion is also proportional to the momentum of such a body times the speed of light ... a product (pc) that gives proportional expression to the area of a square with a side that forms the right side of the aforementioned right angle triangle. Dirac only wanted to work with E and not the quantity E^2 that was entangled in the Pythagorean theorem that was used to give mathematical expression to part of Einstein's theory of relativity because the solutions to the Pythagorean equation indicated there could be both positive and negative results, and before Dirac, no one seemed to know what to make of negative solutions in relation to those sorts of calculations.

Prior to Dirac's work, one of the first advancements that took place with respect to providing a consistent and reliable mathematical means for solving some of the problems surrounding the prediction of future behavioral states entailed by the dynamics or mechanics of quantum entities came from Werner Heisenberg, Pascual Jordan, and Max Born in 1925. They developed a method for describing quantum events through matrices that were capable of representing various properties of particles transitioning over time.

A little later on, in 1926, Erwin Schrödinger introduced his wave equation for describing the manner in which the quantum character of a given physical system evolves with time. Although as far as describing changes in the dynamics of quantum systems over time are concerned, Schrödinger's equation employed a different mathematical method than matrix mechanics did, nonetheless, the two methods are considered to be equivalent to one another ... although, historically, Schrödinger's equation generally has been viewed as being a lot more user friendly than is matrix mechanics.

Schrödinger's equation – at least in its original form – provided a means of describing the evolution of a quantum system only in relation to entities that moved fairly slowly relative to the speed of light. When one used Schrödinger's equation in conjunction with particles moving at high, relativistic velocities (i.e., near the speed of light), his equation delivered nonsensical results. Moreover, Schrödinger's equation did not provide any means of accounting for, or dealing with, the spin of a particle.

On the other hand, Dirac's equation – which surfaced a few years later in 1928 – was capable of dealing with quantum entities traveling at relativistic speeds. Moreover, Dirac's equation also provided a means of taking the so-called intrinsic spin of a particle into consideration in its calculations.

To achieve the foregoing results, Dirac made use of the mathematics of matrices just as Heisenberg, Jordan and Born had done a few years earlier. Nonetheless, the manner in which Dirac employed that form of mathematics was different than the way in which the earlier group proceeded, and in fact it was Dirac's ability to use matrices to keep track of particle rotations in space that became integral to handling the notion of 'spin' in electrons ... something that – as previously indicated – was absent from Schrödinger's equation.

Nonetheless, even though Dirac originally had pursued the mathematics of matrices as a means of – possibly – avoiding the unwelcome negative solutions to various calculations concerning the mechanics or dynamics of quantum entities under relativistic conditions, Dirac found that his intended workaround couldn't achieve its intended purpose. In other words, despite using a different mathematical approach for describing the evolution of a quantum system over time, there still seemed to be a negative dimension to the solutions being generated through his newly discovered equation.

Consequently, Dirac had two choices. He could dismiss the negative solutions as being irrelevant artifacts of the mathematics -- much as his predecessors had been inclined to do with respect to the Pythagorean representation of Einstein's theory of relativity – or, Dirac could try to provide an interpretation concerning the significance of such negative solutions.

Dirac decided to pursue the latter of the foregoing two options. However, his efforts in this regard were not without their problems.

| Quantum Queries |

One possible interpretation of the negative results that arose in conjunction with Dirac's equation -- as well as in relation to other mathematical treatments involving the special theory of relativity -- would be to contend that an electron somehow had assumed a negative energy state. In other words, given that the laws of quantum mechanics indicate that electrons can only take on certain discrete values, and given that the Pauli exclusion principle stipulated that no two electrons can occupy the same quantum state while in close vicinity to one another, then, conceivably, under certain circumstances, electrons might spontaneously drop into a negative energy state.

Unfortunately, such a possibility suggested that matter had the potential to be very unstable since electrons might lower their energy state into a negative mode with ensuing problematic consequences for the dynamics of matter. Moreover, with certain exceptions, matter appears to be fairly stable, and, therefore, one would have to be able to reconcile a possible potential for instability in matter with the fact that the matter that is present all about us appears to be relatively stable.

Dirac's proposal for resolving the foregoing dilemma began with an assumption. More specifically, he proposed that the vacuum was not empty but, instead, consisted of an endless array of negatively energized quantum states that could be occupied by an electron.

The surface of the vacuum would constitute a realm of zero energy. Nonetheless, below that point of zero demarcation were an endless series of negative energy quantum states.

According to Dirac if all of the negative quantum states below the zero point of the vacuum were filled with electrons, there would be no quantum states into which electrons might spontaneously drop. As a result, matter would remain stable.

Furthermore, if one of the quantum states beneath the surface zero point somehow lost its electron occupant, then the absence of such a negatively charged electron would appear as if that quantum state were occupied by a positively charged particle relative to the zero point of energy states. One obvious question that arises in conjunction with the former possibility is how do electrons get dislodged from the negative energy quantum states below the zero point of the vacuum without upsetting the stability of matter.

One possible way of responding to the foregoing question is the following. Let us assume that some high-energy source – such as a cosmic ray or gamma ray – had sufficient energy to enable an electron occupying a negative quantum state in the vacuum to jump to a quantum state above the zero-point.

The presence of the foregoing high-energy source will have caused two things to happen. First, a 'normal' electron would become manifest by being dislodged (through receiving more energy which enables the electron to jump to a higher quantum state from its previous negative energy quantum state in the vacuum), and, secondly, the dislodged electron would leave behind a quantum state in the vacuum that manifests itself as a positively charged electron ... that is, the hole in the vacuum would have quantum properties that behaved like a particle that subsequently came to be known as a positron.

Besides assuming there are negative quantum states that exist in the vacuum which are filled with electrons, Dirac also assumes those states and electrons are endless or infinite in number. Thus, there are actually three assumptions that underlie Dirac's perspective, and any one of those assumptions is vulnerable to issues that raise questions concerning the tenability of his interpretive perspective with respect to the meaning of the negative solutions that arose in conjunction with his equation.

For example, why assume that the negative quantum states said to be inherent in the vacuum are infinite in number? Of course, part of the reason why Dirac presumed things to be the way they were described in the foregoing outline is that by making the number of negative quantum states infinite in character, he wouldn't be confronted with the problem of trying to decide why and how such a series of negative quantum states is finite rather than infinite in nature ... that is, if there is no bottom to the endless series of negative energy quantum states, then one will not have to resolve the problem of what the bottom of the series looks like or why the bottom has the set of characteristics it does.

To assume that the negative energy quantum states of the vacuum are infinite in number is quite arbitrary because there is absolutely no reason why one should suppose this is the case. Nonetheless, for Dirac, being arbitrary in the foregoing manner has the advantage of helping one to avoid having to come up with an equally arbitrary concrete account for why and how the vacuum might be finite in nature since finite considerations are more subject to falsification than are infinite ones ... especially when the infinite often alludes to a hidden and mysterious potential for all manner of strange possibilities ... such as the singularity that supposedly existed prior to the Big Bang.

Another assumption associated with Dirac's position is that quantum states seem to be built into the fabric of nature. Not only are there certain quantum states above the zero-point surface of the vacuum that can be occupied by, say, electrons (states that give expression to the constraints described by quantum mechanics) but, as well, there are certain quantum states that exist below the zero-point surface of the vacuum that do not have to be occupied by electrons in order for those quantum states to take on determinate physical characteristics (e.g., the properties of a positron) in the absence of occupying electrons.

In short, implicit in Dirac's perspective concerning the issue of antiparticles seems to the idea that quantum states have an ontological status independent of the particles that occupy them. Those states determine what particles can and can't do.

Why do the permissible quantum states – whether involving positive or negative levels of energy – have the characteristics they do, and how did they come into existence? Quantum mechanics can mathematically describe certain features concerning those states, but -- to date at least -- quantum mechanics cannot explain why those particular possibilities exist with the properties they have rather than some other set of possibilities.

Finally, all of the negative energy quantum states that exist below the zero-point surface of the vacuum are assumed by Dirac to be occupied by electrons. Apparently, Dirac believes that such an assumption is necessary because, otherwise, electrons might be able to spontaneously drop down into a negative energy quantum state, and, in the process, matter could become unstable.

The foregoing assumption – namely, that electrons tend to occupy the negative energy quantum states below the zero-point surface of the vacuum -- seems unnecessary. Why not suppose that just as there are established quantum states that constrain what an electron can and can't do above the zero-point energy of the surface of the vacuum, there also are established constraints concerning what electrons can and can't do below the zero-point energy of the vacuum.

Why assume there are an infinite number of electrons occupying an infinite number of negative energy quantum states in order to account for the relative stability of observable matter? Why not just suppose that the principles governing quantum behavior prevent electrons from dropping down to energy levels of a negative kind?

In addition, why assume that a given antimatter particle gives expression to the negative energy quantum state that is left behind when that quantum state is vacated by a matter particle? If Dirac is going to go to all the trouble of trying to give an ontological meaning to the negative solutions that arose through the use of his equation to generate intelligible solutions to various problems in physics, then, why not suppose that something – that is a particle of some kind -- is present that has not been taken into consideration by previous models concerning the alleged basic constituents of matter?

During the fourteenth century, William of Ockham is reported to have indicated that one should not multiply assumptions beyond necessity. Yet, Dirac's attempt to account for the negative solutions generated through his equation by assuming an infinity of electrons and an infinity of negative energy quantum states appears to be doing precisely what William of Ockham was advising against ... although, to put things in perspective, William of Ockham's "razor" is a pragmatic suggestion concerning methodology and does not necessarily constitute a law of the universe.

Of course, appreciating the possible relevance of William of Ockham's proposal in relation to Dirac's foregoing interpretation might be considerably easier to do almost 90 years after the latter individual put forth his theoretical account than it was at the time when Dirac came up with his previously discussed idea. Conceivably, assuming that the vacuum was, among other things, an infinite sea of negative energy quantum states occupied by an infinite number of electrons probably made as much sense in 1928 – perhaps more – than assuming that something called antiparticles might have had an ontological status of their own quite independent of dislodged electrons and empty, negative energy quantum states existing below the zero-point of vacuum energy.

However, as far as Dirac's interpretive purposes are concerned, one does not necessarily have to suppose that the vacuum energy is infinite in nature in order for his idea to have operational relevance. For example, one might suppose that the vacuum could be somewhat film-like in nature with a surface that is zero-point energy but with some finite, limited set of negative energy quantum states existing beneath the zero-point surface that are occupied by electrons that are capable of being energized (by, say, a gamma ray) to be able to engage in quantum transitions that take the electron above the zero-point energy of the vacuum while leaving behind a positively charged electron-like hole.

As long as the film-like vacuum contained a sufficient number of vacuum electrons that could play musical chairs with the negative energy quantum states from which electrons might be dislodged and, therefore, be able to provide electrons as needed for different negative energy quantum states in order to be able to prevent non-vacuum electrons from dropping down to negative energy states in the vacuum, then Dirac's idea still might work. Of course, what would constitute a "sufficient number of vacuum electrons" to ensure the stability of matter is unknown, but the point being made here is that one wouldn't necessarily need an infinite number of such vacuum electrons or an infinite number of negative energy quantum states in order for Dirac's idea to make operational sense.

From the perspective of the foregoing modified edition of Dirac's original idea, the vacuum still would be the source of a certain amount of energy that, among other things, would be able to help stabilize matter. However, the amount of energy in such a vacuum need not be, and might not be, infinite in character.

Given that physicists in 1928 already were confronted with embarrassing problems involving unwanted infinities (e.g., the charge and mass of allegedly point particles like the electron), one might have thought that Dirac would have been somewhat more circumspect about introducing several additional kinds of infinity into the theoretical mix – especially if (as indicated above) such a move might not be necessary. Nonetheless, as indicated previously, assuming something is infinite in nature tends to hide a lot of questions – at least mathematically.

The real problem comes when one tries to give physical meaning to such infinity-friendly mathematics. One can no longer conceal the many questions that arise in conjunction with the concrete and the finite beneath the mysterious and alluring – but vague and amorphous -- veils of the infinite.

There is another difficulty entailed by Dirac's interpretation involving the possible significance of certain negative mathematical results that arose from using his equation to calculate solutions to various problems in physics. This issue would not have occurred to Dirac in 1928 since at that time no one knew that matter and its antimatter partner annihilated one another because prior to Dirac no one suspected that antimatter might actually exist.

More specifically, to hypothesize that electrons occupied negative energy quantum states that acted like positively charged electrons when the occupying electrons were dislodged from those states tends to be somewhat problematic ... at least, it would become so in the light of the results of subsequent research. After all, many later experiments would demonstrate that electrons and their antiparticle partners – positrons – cannot coexist for very long, so, given the reality of annihilations that take place when matter and antimatter consort with one another, then, seemingly, electrons would not have been able to occupy negative energy quantum states that acted like positively charged electrons in the absence of such electrons in the way that Dirac might have envisioned.

In modern terms, Dirac's idea concerning the nature of the vacuum appeared to suggest that matter and antimatter could be coupled together and nothing much would happen. Dirac, of course, was not really saying this but, instead, he was trying to provide an explanation for some of the mathematical results that arose through the use of his equation.

However, if Dirac's original idea is to survive the experimentally demonstrated fact that matter and antimatter cannot coexist together without explosive results, then, obviously, there would have to be some sort of modulating dynamic. Such a dynamic would need to permit a certain degree of coexistence between electrons and the negative energy quantum states below the zero-point surface level of the vacuum they occupied.

The efficiency of matter-antimatter annihilations is total. Although chemical reactions release about one billionth of the energy stored in atoms, and while fission reactions release about 1/1,000 of the energy present in atoms, and whereas fusion reactions unlock about 1% of the energy in atoms, all of the energy present in particle-antiparticle annihilation reactions is released.

What actually takes place during the dynamic of annihilation? We know some of the features of that dynamic, but there appear to be some missing pieces.

We know there seems to be some sort of interactional dance that takes place between, at least, certain kinds of matter and antimatter pairings (e.g., protons and antiprotons) before the final consummation of annihilation occurs. Nevertheless, the reasons why that dance transitions into total, mutual annihilation are shrouded in mystery.

Why and how does a difference in charge lead to annihilation rather than, say, a bond of some kind? Why do some charges of an opposite nature attract (e.g., electrons and protons) while other charges of an opposite nature (electrons and positrons) lead to annihilation?

Are there differences in the nature of the charges involved in the foregoing cases? Are there differences in the role those charges play in matter-antimatter pairings but do not play in matter-matter pairings involving oppositely charged particles?

Perhaps electrons might be able to occupy Dirac's proposed negative energy quantum states -- that manifest some of the properties of a positively charged electron when the occupying electron is dislodged --because some aspect of the potential for annihilation is temporarily blocked during those kinds of occupations so that the dance can't complete itself (somewhat like the way molecules that attach to membrane proteins can alter and restrict the normal action of those membrane proteins during the process of competitive inhibition). To hypothesize that an electron occupies a negative energy quantum state doesn't really specify how that electron occupies such a state.

Perhaps, if an electron were to occupy such a negative energy state in one way, the pre-annihilation dance doesn't take place or isn't able to complete itself. On the other hand, if that occupation occurs in a slightly different way, then, annihilation occurs.

We really don't know much about the nature of the vacuum. There might be more, or less, going on in the vacuum than Dirac assumed to be the case.

Einstein, via $E=mc^2$, showed that matter is merely energy that has been constrained in a particular way. Dirac, via his 1928 equation: $i\gamma \cdot \partial \psi = m\psi$, indicated that when matter arises through the dynamics of energy, one of the by-products is a phenomenon that later came to be referred to as antimatter ... a notion that lent intelligibility to the negative solutions that arose in the context of both Dirac's equation as well as the special theory of relativity.

As outlined previously, Dirac's initial theoretical attempt to explain those mirror-like reflections of "normal" matter appeared to entail various kinds of questions and problems. Nonetheless, his idea constituted an important change in conceptual direction that would begin to pay concrete dividends when Carl Anderson experimentally verified the existence of the positron in 1932.

However, neither Einstein nor Dirac could explain how energy came to be constrained in the form of a particular collection of quantum properties involving matter and antimatter. To be sure, the quantum properties that emerged from a given context of energy were the most stable arrangements possible relative to such a context, and, in addition, the nature of the particles that arose within such a context were constrained by the amount of energy available, and were constrained, as well, by various laws of conservation.

Nonetheless, the precise nature of the conversion of energy to matter and antimatter remains something of a mystery. How does energy -- say in the form of photons (but there are more than a hundred possible forms other than photons that could be substituted here and give rise to the same question) -- transition into specific quantum states involving, say, electrons and positrons? What orders or organizes the re-configuration of photons (or other mediums entailing such transitions) in a way that generates the constraints and structure that constitute particles with mass and certain quantum states?

In other words, how does the dimension of order (e.g., conservation laws, thermodynamic issues of stability, quantum properties) become part of the process of transition that generates matter and antimatter of a particular kind from a given context of energy? Quantum mechanics is able to describe the quantitative aspects of such a transition in a very precise way, but quantum mechanics can neither describe nor explain the nature of the dynamics through which the ordered dimensions of matter and antimatter arise out of energy.

Given a set of starting conditions, quantum mechanics can tell one what the outcome is likely to be if those starting conditions are permitted to give expression to their potential. However, quantum mechanics cannot tell one what actually takes place during the transition from starting conditions to quantum outcomes.

As noted earlier, Dirac's theoretical account -- concerning the meaning or significance of certain kinds of negative solutions that arose in conjunction with his equation -- held that there are negative energy quantum states beneath the surface of the zero-energy point of the vacuum ... states that exist even after electrons have been dislodged from them. Is it possible that the universe (including the vacuum) is filled with (or made from) those sorts of quantum states and that the states become realized when the right amount of energy is available to activate those quantum states?

Maybe, fields (of whatever nature) all give expression to a set of variable quantum states … potentials that are realized in the presence of appropriate conditions of energy. If so, then, where and how quantum states with such properties arose -- that is, how fields come to have the ordered constraints they do -- becomes an important issue.

In any event, several years later -- during the month of September 1931 – Dirac reformulated his theoretical position somewhat in order to avoid a variety of problematic questions concerning his views about the possible meaning or significance of some of the peculiar mathematical results associated with use of his equation. He proposed the existence of a new, but not yet discovered, particle and referred to the particle as an anti-electron.

In fact, his re-configured perspective of 1931 stipulated there was a symmetry existing between negative and positively charged particles. Not only did every particle have an antiparticle counterpart, but, as well, Dirac maintained that if the symmetry linking matter and antimatter were truly fundamental in character, then, under the right circumstances, the charge of any particle could be reversed.

Despite the fact that Dirac altered the character of his original idea, the idea of a non-empty vacuum remained. For example, Dirac believed that not only do electrons occupy the infinite sea of the vacuum – as noted earlier – but, as well, protons also are contained in that sea.

Consequently, the problems and questions that have been raised previously in this chapter with respect to the infinite character of the vacuum in relation to electrons also apply to the idea that the vacuum contains an infinite supply of protons. I believe a more tenable position concerning the nature of the vacuum might be that if the vacuum does contain some amount of energy and/or set of quantum states, then those amounts are finite in character rather than being infinite in character.

| Quantum Queries |

The alleged presence of infinities in nature always seems to lead to difficulties in science. Indeed, Planck initiated the quantum revolution by proposing – rather arbitrarily at the time -- that there was a finite solution – i.e., the Planck constant – to the problem of infinities associated with calculating black body radiation for certain frequencies.

Something similar might also be true with respect to working out the properties of the vacuum. In fact, at a certain point, problems concerning the nature of the vacuum tend to merge with problems concerning the nature of space ... a topic to which I will return in the next chapter.

Currently, the so-called Standard Model of quantum physics is the most up-to-date way to describe the manner in which the forces and particles of nature interact with another. Although various kinds of refinements to the model continue to take place, there are relatively few instances of physical phenomena involving particles that can't be described and, to a certain extent, explained through the use of the mathematical and experimental wherewithal of the Standard Model.

One exception to the foregoing general statement involves neutrinos. At the present time, physicists are not quite sure how to incorporate neutrinos into the Standard Model but have proposed extending that model through various means (e.g., grand unified theories and Supersymmetry models ... both of which will be touched upon shortly).

When Wolfgang Pauli came up with his idea of the neutrino in 1930 he was not trying to hypothesize the existence of some major player in the material realm. Instead, he was merely trying to imagine what might account for a tiny bit of energy that appeared to go missing in conjunction with certain kinds of particle interactions ... an entity -- if it existed – that would have to be so small that Pauli feared no one would ever be able to detect its presence.

Some individuals have estimated that the number of neutrinos in the universe is somewhere in the vicinity of 10^{89}. No one actually has verified whether, or not, such an estimate is correct.

Estimates like the foregoing one are usually functionally dependent on a variety of assumptions about how the universe came into being and how it operates. If there are substantial problems entailed by any of those assumptions, then, the foregoing figure might have to be re-calculated to some degree – either in an upward or downward direction.

Knowing how many neutrinos exist, as well as knowing what the mass (masses) of those neutrinos are, involve issues that could play influential roles with respect to being able to construct a viable theory of cosmology concerning the origins of the universe. Individually, neutrinos might appear to be trifling entities, but collectively, some theorists believe that, over time, they might have had the capacity to play an important role in helping to shape various aspects of the structure of the universe.

Neutrinos come in at least three varieties – tau neutrino, muon neutrino, and electron neutrino. A fourth variety – known as a "sterile neutrino" has been hypothesized to exist, and if the latter neutrino does exist, it might not be subject to the weak force as other neutrinos are (and, hence, the designation "sterile").

For reasons unknown to modern science – at least for the moment – neutrinos seem to have the capacity to switch from one variety of neutrino to another. The capacity to switch flavors or oscillate depends, to some extent, on properties called 'mixing angles' that help set the parameters within which such oscillations occur.

The mass of any given neutrino might consist of different flavors of neutrino. Conceivably, through some kind of oscillation mechanism, the manifested characteristics of a neutrino might alternate as a function of the mixing angles of the differently flavored neutrinos that make up the mass of a specific electron, tau, or muon neutrino.

However, the possibility that neutrinos are able to switch flavors also appears to depend on the still unresolved issue of the mass of the different kinds of neutrino. According to the principles of quantum mechanics, neutrinos have the capacity to switch identities with one another only if those neutrinos have mass and only if that mass is different in each instance.

Currently, physicists know that at least two of the three forms of neutrinos have a non-zero mass. Beyond that, however, not much is known with any precision.

For example, no one knows which of the three neutrinos is heaviest or lightest or in between. One of the values used in conjunction with the estimated mass of a neutrino is the quantity: 0.00000000000000000000000000000000018 kg.

However, the foregoing figure does not so much constitute the actual mass of a neutrino as it constitutes an upper limit on the possible size of a neutrino. Therefore, the mass of two of the three known neutrinos might actually be somewhat less than the foregoing upper limit indicates, but no one knows how much less the actual mass of any those neutrinos might be than the foregoing figure suggests could be the case.

Neutrinos are estimated to have anywhere from one millionth of the mass of an electron (the heaviest neutrino) to one hundred millionth of the mass of an electron (the lightest neutrino). Whatever the precise mass of the neutrino might be, scientists are not exactly sure why neutrinos are so light.

The Standard Model of particle physics predicts that the neutrino is massless. Given that at least two of the three known kinds of neutrinos are not massless, the Standard Model would seem to be incomplete in some way.

Neutrinos are the only known particles that are electrically neutral. Neutrons -- which are electrically neutral in a holistic sense -- consist of charged quarks, and while the overall electrical character of neutrons is neutral, this is only because of the way in which the differential charge dynamics of the neutron's internal quark arrangement counterbalance one another.

Conceivably, neutrinos – like neutrons -- might also have some sort of a complex interior life in which an inner dynamic involving various kinds of electrical charges gives expression to an electrically neutral whole. However, if neutrinos do have a complex interior, then due to current limits involving technical/methodological capabilities, scientists might be a long, long, long way from being able to demonstrate that neutrinos have an inner structure.

Because neutrinos are electrically neutral, they are not affected – at least directly -- by electromagnetic fields. Moreover, like other leptons – such as the electron -- neutrinos are not affected by the strong force.

Being relatively immune to both electromagnetic fields and the strong force, neutrinos do not appear to play much of a role in the dynamics of atomic structure or the phenomena of chemistry. Nevertheless, neutrinos are generated during radioactive decay, and, therefore, neutrinos are entangled in the dynamics of the weak force.

Neutrinos possess a property known as handedness. Such handedness is tied to the presence of the weak force.

To date, no right-handed neutrinos are known to exist, and, therefore, all three known modalities of neutrinos are left-handed. Yet, if such right-handed entities did exist, this property might constitute the means through which neutrinos could be incorporated into the Standard Model.

If there were actual right-handed forms of neutrinos, then, the existence of such handedness in neutrinos might be able to help explain why neutrinos are so light. However, if neutrinos with such handedness do exist, they will be more difficult to discover than their left-handed brethren have been ... and the latter discoveries have not been all that easy to accomplish.

The relatively recent announcement concerning the existence of a Higgs field carries implications that could affect left-handed neutrinos but not necessarily right-handed ones (if the latter particles actually exist). Consequently, if right-handed neutrinos do exist, the source of their mass presumably might be generated through some sort of non-Higgs mechanism.

While trying to account for how the universe came to be the way it is, some so-called grand unified models believe that enormously high energies were operative at one time in the universe (prior to the Big Bang), and those grand unified models predict that right-handed neutrinos would have been fairly heavy within such a high-energy context. Some of those grand unified models contend that there might be some sort of unknown quantum process which would permit left- and right-handed neutrinos to interact in a way that would give expression to a left-handed neutrino that was very light.

In other words, by means of the sort of unknown, quantum interaction being alluded to through such grand unified theories, the hypothetical heavier, right-handed neutrino would help mask or modulate the mass of the left-handed neutrinos in some fashion. In the process, left-handed neutrinos might show up as very light particles.

The foregoing process is referred to as the 'seesaw mechanism'. Somewhat like a seesaw, the higher mass of the hypothetical right-handed neutrino has an impact on how the mass of left-handed neutrinos will manifest themselves.

Grand unified theories revolve about the idea that in the high-energy environment that is hypothesized to have existed prior to the Big Bang, the forces of nature (with the exception of gravity) formed one symmetry group and, then, became differentiated into strong, weak, and electromagnetic forces when the initial symmetry of unification was broken somehow.

The energies envisioned by such models are significantly beyond the capabilities of current accelerators, and, consequently, since those energies cannot be recreated in the lab, whatever evidence is to be gathered in support of that kind of a theory would have to be indirect in nature.

Grand unified theories predict an array of possibilities that might serve as indirect evidence that such theories are correct. For example, those theories predict that protons are susceptible to decay.

Although a great deal of effort and time has been spent trying to detect the predicted products of proton decay, so far, there is no evidence that protons actually do decay. Moreover, if evidence consistent with proton decay should continue to go missing in action, then, this could constitute evidence that grand unified theories – at least in their current forms – are not viable accounts of how the universe came to be the way it is.

In addition, various versions of grand unification theory predict the existence of magnetic monopoles. As has been the case with respect to the matter of proton decay, to date, there is no evidence indicating that magnetic monopoles actually exist, and, therefore, the editions of grand unified theories that predict the existence of magnetic monopoles entail yet another kind of evidential weakness – at least for the time being -- with respect to the issue of scientific viability.

An alternative to grand unified theories is a model known as Supersymmetry. Like grand unified theories, Supersymmetry models also try to account for, among other things, the light masses of neutrinos.

Some versions of Supersymmetry already have been put to the test at CERN in conjunction with the experiments that was conducted over a period of years -- ending in 2013 -- of the LHC (Large Hadron Collider ... a hadron is any particle that is sensitive to the presence of strong force dynamics ... for instance, protons, neutrons, and mesons are hadrons). Such versions of Supersymmetry required the Higgs Boson to have a certain mass if those theories were to produce viable results, and, consequently, when the experimentally determined size of the Higgs Boson – around 125-126 GeV (i.e., billion electron volts) – was established in 2012, a number of Supersymmetry models were eliminated from consideration because they were based on hypotheses that the Higgs boson had a larger mass than it was experimentally shown to have.

Other editions of Supersymmetry involving a Higgs boson that better reflects the experimental result established in 2012 will be put to the test in the upcoming series of experiments that will take place beginning some time in early 2015 at the LHC in CERN. It remains to be seen what the status of Supersymmetry models will be following those experiments (This issue is discussed a little further in the final chapter of this volume).

Supersymmetry predicts the existence of a variety of particles – sometimes referred to as 'superpartners' -- that, so far, have not been shown to exist. Indeed, many physicists hope that telltale signs of such Supersymmetry particles will show up in the data that will be generated over the next several years at CERN's LHC.

Nonetheless, some particle physicists have become somewhat nervous about the viability of Supersymmetry. Those individuals feel that if such 'super' particles do exist, then indications of their existence should have shown up already during the last set of experimental runs conducted at CERN, and this did not appear to have happened.

If the hypothesized super particles do not show up in the data during the forthcoming set of experiments to be run at CERN beginning in the spring of 2015, this sort of absence of evidence with respect to such particles might be considered by some individuals to constitute evidence of absence with respect to the hypothesized existence of those super particles. Moreover, if such super partners do not exist, then Supersymmetry will not be giving expression to the sort of explanatory model that many physicists hoped would be the case ... indeed, that model might just be wrong.

According to Supersymmetry, every known particle has a 'super' counterpart. A 'super' particle that is associated with neutrinos is known as a "neutralino".

The stability of this hypothetical particle depends on whether, or not, it (along with all other proposed super particles) possesses a hypothetical property that is referred to as 'R-parity' which – if it exists – prevents super particles from decaying into the particles that currently are encompassed by the Standard Model.

If the hypothetical neutralino does not possess the hypothetical property of R-parity, the neutralino would be unstable and subject to decay. The precise nature of that decay would depend, to some extent, on the mass of the neutrino – for which an upper limit has been established but whose actual nature is not, yet, known.

In addition, the mass of the neutrino depends on whether one is talking in terms of left-handed (which actually exist) or right-handed (hypothetical) versions of a neutrino. Left-handed neutrinos might operate in accordance with a different mechanism for generating mass than their hypothetical right-handed counterparts do.

Another facet of the possible relation between neutralinos and neutrinos revolves around the issue of whether, or not, neutrinos are their own antiparticle. Particles that are their own antiparticles are known as Marjorana particles, whereas particles that are not their own antiparticles are referred to as Dirac particles.

According to some models of Supersymmetry, hypothetical right-handed neutrinos could affect the way left-handed neutrinos express their mass (that is, in terms of being extremely light) not only through the (currently unknown) mechanism that supposedly gives right-handed neutrinos their large mass but, as well, through the properties inherent in the dynamics that are theoretically predicted if right-handed neutrinos are their own antiparticles.

The class of quantum entities known as leptons (electron, tau, and muon particles plus their respective neutrino particles) has been observed to conserve the property known as "lepton number". That is, during any given particle dynamic, the number of leptons minus the number of antileptons is a conserved quantity and, therefore, stays the same across such a dynamic.

Conceivably, however, through the presence of neutralinos and right-handed neutrinos that are their own antiparticles (and all three of the foregoing features are hypothetical in nature), the aforementioned property of lepton numbers might not be conserved. If this were to be the case, then an excess of matter over antimatter might arise (which brings us back to the topic with which this chapter began).

Even if turns out to be true that: (1) neutralinos exist; (2) right-handed neutrinos exist; (3) right-handed neutrinos are their own antiparticles, and (4) all of the foregoing hypothetical entities – at this point anyway – have the characteristics they are calculated to have by one, or another, model of Supersymmetry, a further issue still remains. Will the hypothesized existence of an excess of matter over antimatter be the sort of quantity that can account for why only matter seems to exist in the universe and why antimatter only seems to arise as a sort of second-fiddle to, and is largely constrained by, the dynamics of matter? Or, stated from the perspective with which this section on Antimatter began, will the foregoing four hypothesized properties result in an excess of one unit of matter for every 10 billion pairings of matter and antimatter that has been calculated by some individuals to have been necessary to lead to the universe we see today?

With respect to (3) above – that is, the issue of neutrinos constituting their own form of antimatter (remember, neutrinos are electrically neutral and, therefore, there is no neutrino antiparticle which will have an electrical charge that is opposite from a neutrino) – a number of experiments have been carried out to test the foregoing hypothesis. Those experiments involve a phenomenon known as double beta decay.

Single beta decay occurs when a radioactive nucleus becomes more stable after a neutron is transformed into a proton and, in the process, an electron and an antineutrino are radiated. For example, when the neutron in tritium – an isotope of hydrogen – transmutes into a proton through radioactive decay, helium 3 results, and, during that transformation, an electron and an antineutrino arise as by-products.

There are a few radioactive nuclei – for, example, Germanium 76 – that are capable of undergoing double beta decay. This occurs when two neutrons are transmuted into two protons and such a transmutation gives rise to a more stable atom (Selenium 76 in the case of Germanium 76) and, as well, two electrons and two antineutrinos are generated as by-products of that transmutation.

If neutrinos are their own antiparticles, physicists have predicted, then the two aforementioned antineutrinos should be able to cancel one another. For instance, a neutrino might be absorbed rather than an antineutrino being radiated.

Two electrons (leptons) would have been radiated but counterbalancing antineutrinos (antileptons) would not have emerged during such a process. In other words, if the foregoing scenario were correct, then, certain kinds of double beta decay might violate the conservation principle that normally holds sway in conjunction with lepton particle numbers.

The general process – if, for example, double beta decay actually occurs – is known as leptogenesis. Leptogenesis – if true – might help to explain at least some of the asymmetry between matter and antimatter that is observed in the universe because the process generates an excess of leptons relative to antileptons.

To date, the jury is still out with respect to the foregoing possibilities. The data concerning double beta decay phenomena are inconclusive with respect to the issue of whether, or not, neutrinos serve as their own antiparticles.

Even if indisputable evidence were forthcoming indicating that neutrinos were their own antiparticles and, as a result, lepton number is not always conserved, nonetheless, the path to establishing a plausible account of the existing asymmetry between matter and antimatter would not necessarily be straightforward. The asymmetry between matter and antimatter is across the board ... in other words the asymmetry goes beyond leptons and extends into all manner of matter-antimatter pairings.

Of course, _if_ right-handed neutrinos were to exist, and _if_ neutralinos were to exist, and _if_ R-parity did not hold with respect to such neutralinos, and _if_ unstable neutralinos were to have the right kind of decay products, then, one might be able to introduce a bit more asymmetry into the picture with respect to the observed character of the relationship between matter and antimatter in the universe. However, the foregoing scenario is not only highly speculative, but, as well, there is no guarantee that even if everything in the foregoing scenario that is hypothesized to be true were, in fact, true, nevertheless, that result still would not necessarily be able to account for why things are the way they are with respect to the general asymmetry between matter and antimatter of all kinds.

The asymmetry involving matter and antimatter appears to be a function of something more than just the number of different kinds of particles that are in existence. That asymmetry seems to be inherent in the way the universe operates.

With the possible exception of certain relatively specialized cases during which, say, the principle of lepton conservation could be violated (e.g., if neutrinos were their own antiparticles), nonetheless, for the most part, the tendency of particle dynamics seems to be inclined toward preserving principles of conservation rather than violating those principles. Moreover, to whatever extent such principles of conservation are violated, then, perhaps, it might not only be the case that those exceptions are insufficiently great to disrupt the attractor basin that gives expression to universal dynamics, but, as well, one might also suppose that those sorts of anomalies or perturbations would be constrained and/or dampened through the manner in which the universe normally operates.

Antimatter does not tend to show up unless the condition of equilibrium is disturbed through a dynamic that involves a redistribution of energy and particles. The tendency of matter is to move toward a state of equilibrium and dampen the presence of antimatter -- when it does arise -- through paired annihilations.

The generation of antimatter is a means of re-distributing energy in a stable way through the annihilations that take place. This mode of energy distribution is an expression of the Hamiltonian -- the way the energy system operates as a whole – and, as such, it suggests that the Hamiltonian has an inherent bias against the generation of the sort of antimatter that has the capacity to linger on in a relatively stable form as most matter does.

The aforementioned Hamiltonian bias gives expression to what might be referred to as a Planckian constraint concerning the nature of dynamics in general. Just as Planck arrived at the conclusion – in a move of desperation -- that there must be a finite and constant quantum of action constraining how energy was emitted or absorbed in black body dynamics in order to be able to make sense of the available experimental data, so too, there are other constraints that appear to be built into the way the universe operates ... such as the Hamiltonian bias involving the means through which, and the way in which, certain facets of energy are redistributed by means of matter and antimatter pairings in order to help stabilize the system as a whole, and in the process, help make sense of the observed asymmetry existing between matter and antimatter.

Another kind of Planckian constraint involves the weak force. For example, how do the atoms in a radioactive isotope "know" how many atoms constitute half of the remaining atoms in such a substance?

One possibility, of course is that radioactive isotopes don't 'know' how many atoms have to decay in order to comply with the half-life law governing that isotope. Rather, just as there are pre-existing "orbitals" that set quantum conditions that constrain how electrons can interact with protons within an atom, so too, there might be pre-existing quantum conditions associated with a given isotope that constrain which atoms in a given isotope will, and will not, be subject to the action of the weak force and, as well, there might be conditions associated with a given isotope that constrain or set the quantum conditions that determine the rate of decay that is peculiar to such an isotope.

The 'isotope decay repository' (counterpart to the idea of orbitals for electrons) is filled with all those atoms that are not, yet, susceptible to the effect of the weak force, just as the orbitals of an atom are filled with all those electrons that meet the conditions of that orbital. The "switch" or quantum property that renders the atom of a radioactive isotope temporarily immune to the weak force is turned to an "on" or "off" position (like the 'up' or 'down' spin of electrons), and wherever such atoms exist in the isotope, then just as the quantum state of an electron determines its membership in a given orbital irrespective of where that electron might be in the atom, so too, those atoms with the requisite quantum properties that render them temporarily immune to the weak force gives expression to de facto membership in the isotope decay repository that can hold only so many atoms -- which turns out to be half of what remains – just as orbitals come with pre-established constraints on how many electrons can occupy such an orbit.

The half-life is the pre-existing quantum constraint (or set of such constraints) placed on radioactive substances that constitutes the most stable way of proceeding under a given set of unstable circumstances. The half-life of a substance is its manner of ratcheting down to more stable states via established levels of constraint just as different orbitals establish stable conditions for constraining the transfer of energy/particles either in a more energetic or less energetic direction.

In addition, just as different atoms have different sets of orbitals that help give expression to some of the properties (e.g., chemical) of those atoms that are unique to them, so too, different isotopes have different settings for how radioactive decay expresses itself over time ... a mode of decay that tends to be unique for different isotopes and which is a function of the pre-existing quantum conditions governing or constraining how the weak force can operate in conjunction with such isotopes.

Irrespective of whether one is discussing Hamiltonian stability in relation to matter and antimatter asymmetry, or one is speaking about the constant of quantum action, or one is talking about the half-life of radioactive isotopes, then in each of the foregoing cases one is alluding to the same kind of phenomenon. These are all expressions of Planckian constraints that appear to be inherent in the structure of the universe that govern the conditions within which certain kinds of dynamics take place.

If the foregoing perspective is correct, then, there is not necessarily any need to explain how the asymmetry between matter and antimatter arose because what is currently observed in this respect is the way things always have been. Whether one wishes to promote the idea of a Big Bang or a Steady State universe, it is possible that the present asymmetry between matter and antimatter has always been inherent in the dynamics of the universe and, consequently, there has never been a cosmic juncture during which matter and antimatter were in a condition of thermal equilibrium.

Conceivably, one should not be looking for a model that can explain how the asymmetry between matter and antimatter arose. Possibly, one should be looking for a cosmological model that comes with a built-in bias, as it were, toward a set of dynamics in which such asymmetry is already present.

To be sure, the latter possibility is as speculative as any of the other possibilities (e.g., grand unified models, Supersymmetry, and Dirac's model) that have been explored somewhat throughout this chapter. On the other hand, the idea that the universe is, for example, inherently biased toward an arrangement that is dominated by matter and in which antimatter plays a largely secondary role is a much simpler theory than the ones being proposed through grand unification and Supersymmetry.

Even if some version of Supersymmetry or grand unification theory proves to be true, both of them actually constitute models that have an inherent bias concerning the relationship between matter and antimatter. So, to date, all of the possibilities that have been introduced thus far in order to account for the asymmetry between matter and antimatter just give expression to different ways of arriving at the fact on which they all agree – namely, asymmetry -- because that is what is observed in the universe today.

Nonetheless, at the present time, after all is said and done, the reason (or reasons) why there is asymmetry between matter and antimatter remains a mystery. Moreover, there seems to be no more reason for favoring any one of the previously mentioned theoretical possibilities over any of the other models that have been touched upon with respect to the issue of accounting for the origins of the asymmetry between matter and antimatter that is observed in the universe.

The forthcoming rounds of experimentation that are to take place in conjunction with the LHC at CERN might, or might not, provide the sort of evidence that will help resolve the current mystery concerning matter-antimatter asymmetry. On the other hand, such experiments might generate data that eliminates various models from consideration with respect to -- among other things -- the matter-antimatter issue.

Chapter 3: All Tangled Up

Erwin Schrödinger introduced the term "entanglement" in an article that appeared in the *Proceedings of the Cambridge Philosophical Society*. It refers to a phenomenon in which quantum entities that once were engaged in a dynamic of some kind are able to continue to be sensitive to certain modalities of change taking place with respect to one another despite having become separated from each other by considerable spatial distances.

One of the reasons – but not the only one – that stood in the way of Einstein being able to whole-heartedly endorse quantum theory (despite his acknowledgement of its heuristic value) involved the issue of spooky actions at a distance that seemed to be entailed by quantum dynamics. He wanted physical events to take place within a framework of time and space that wasn't haunted by ghostly interactions among quantum particles that apparently were able to operate contrary to the way in which Einstein believed reality worked.

More specifically, Einstein maintained that events are constrained by the condition of locality. He believed that in order for one object to affect another, there had to be some form of direct contact involving those objects and through which they were causally affected.

Furthermore, Einstein believed that any mode of contact that acted upon some object took time to occur. Therefore, according to Einstein, one object could only affect another object if the two objects were linked through conditions of locality involving time and space that could account for how an effect arose out of a cause when two objects were contiguously connected, in some dynamic fashion, via time and space.

Entanglement, however, appeared to violate the condition of locality. The interaction between two objects said to be entangled seemed to be instantaneous and, therefore, of a nonlocal character.

In other words, the phenomenon of entanglement did not appear to require any traversing of a distance in space that took some period of time to complete. This idea of nonlocality was the spooky and ghostly action at a distance to which Einstein was referring in, among other places, the position developed by him as well as Podolsky and Rosen in 1935.

The action at a distance problem had previously raised its mysterious head in 1688 when Newton stated in his work, Principia *Mathematica* that while the power of gravity was: Real, observable, capable of being quantified, and law-like in its behavior, nonetheless, he didn't have any idea of how gravity was propagated between objects. He stipulated in his book that he would not frame any hypothesis (hypotheses non fingo) concerning how gravity did what it did, but however gravity worked Newton assumed that the phenomenon operated instantaneously.

In his general theory of relativity, Einstein removed the spooky action at a distance dimension of Newton's theory that was implicit in the notion of instantaneousness. Instead, Einstein believed that gravity traveled at the speed of light, and, therefore, gravity like everything else in the universe was subject to the condition of locality.

Although Einstein's assumption concerning the speed with which gravity propagated its effects was subsequently proven to be true, mysteries concerning the nature of gravity continue to this day. Einstein's general theory of relativity described gravitational phenomena in terms of geometry, and, indeed, he said that gravitation is geometry, but nonetheless, the actual nature of gravity continues to elude the scientists of today as much as it's nature eluded Newton in his day and Einstein in his time.

The issue of nonlocality was not the only problem Einstein had with quantum theory. He also took exception with the way many physicists apparently were willing to doom reality to being nothing more than a function of statistics and probabilities that were independent of any kind of an underlying reality – sometimes referred to as hidden variables – that were responsible for generating and shaping the physical dynamics that were being described through the statistics and probabilities that came to be associated with quantum events.

Einstein was perfectly willing to use the mathematical techniques of statistical mechanics and quantum probabilities to describe various physical phenomena – such as in relation to Brownian movement. However, Einstein believed there was some sort of physical dynamic taking place that was being described, within limits, through such mathematical techniques.

For example, initially, Max Planck treated his notion of quanta as being merely a sort of arbitrary mathematical construction that could help reconcile mathematical descriptions of black body radiation with experimentally determined results. He didn't consider quanta to be real, and was critical of Einstein when the latter individual claimed that light consisted of Planck's imaginary quanta and could be used to account for why a small current of electricity would be produced when light was directed toward certain metals and, in the process, gave expression to the photoelectric effect.

Not only did Einstein's account of the photoelectric effect treat Planck's mathematical artifact as real, but, as well, Einstein was throwing down the gauntlet with respect to a long-standing understanding – established through the interference experiments run by Thomas Young in 1801 -- that light propagated as a wave. According to Einstein, light propagated in the form of discrete packages of energy or quanta.

In 1924, Louis de Broglie appeared to re-introduce the relevance of wave descriptions with respect to quantum phenomena. De Broglie hypothesized that there is a wavelength, λ, which can be linked with any particle or object that is related to the latter's momentum, p, via the Planck constant, h.

Conceivably, Planck, Einstein, and de Broglie were all both correct and incorrect at the same time. Perhaps, one needs to make a distinction between the mathematics that is used to describe some given phenomenon and the reality to which such mathematics alludes.

For instance, are quantum phenomena wave-like or particle-like? Many experiments over the years have demonstrated that under various conditions quantum events give expression to phenomena that have properties that can be reliably described through the mathematics of waves as well as through the mathematics of quanta.

Some scientists have referred to the presence of the foregoing sort of duality as an indication that quantum phenomena are both wave-like and particle-like. The term "wavicles" is sometimes used to refer to that sort of duality.

However, one might also argue that reality is not necessarily either wave-like or particle-like, but, instead, gives expression to something more fundamental than quantum events ... something that has the capacity to manifest wave-like and particle-like properties under the right conditions. As such, quantum events constitute an intermediate medium of sorts that operates in accordance with a set of principles that, depending on circumstances, has the capacity to generate either wave-like or particle-like phenomena even as those principles are neither necessarily wave-like nor particle-like in character.

If this is the case, then, maybe, one should be careful to distinguish between, on the one hand, the inherent nature of reality, and on the other hand, the quantum properties that arise as a result of the former dynamic. If the reality being alluded to through the mathematics of quantum mechanics is treated as that which makes a given set of properties possible and that, under different circumstances, can be accurately described through both wave-like (e.g., de Broglie and **Schrödinger) and particle-like treatments (e.g., Einstein and Compton)**, then, the mathematics does not necessarily fix or determine what such events actually are. Rather, the mathematics merely describes some of the phenomena to which the underlying reality – whatever it might be -- gives rise.

In other words, perhaps, quantum events are neither a function of waves nor a function of particles. The reality underlying quantum events is something that can give rise to phenomena that are both particle-like and wave-like in character but that "something" cannot be reduced to being either wave-like or particle-like in nature.

If one were to adopt the perspective that has been outlined in the last several paragraphs, then, one might come to understand how Planck, Einstein, and de Broglie could be considered to be simultaneously both correct and incorrect. For example, Planck might have been correct to consider the notion of quanta as merely being a mathematical construction that helped solve some problems involving black body radiation, but, at the same time, Planck also might have been wrong to want to claim that quanta were not tied, in a rather complex manner, to an underlying reality that constrained the value of Planck's quantum of action to be a constant of a given kind (i.e., **$6.62606957 \times 10^{-34}$ m² kg / s**) ... in other words, quanta were not real in any fundamental sense, but the nature of that which constrained the phenomena to which the term "quanta" alluded was real and not an arbitrary artifact of his mathematical calculations.

Similarly, Einstein might have been correct to indicate that when light is directed toward certain kinds of metals, then, one will observe a small electric current that can best be described as a particle-like phenomena. On the other hand, Einstein also might have been wrong to suppose that the underlying reality that was making such a particle-like phenomena possible was, strictly speaking, a function of quanta ... but, instead, dynamics were being constrained and ordered in such a way that made reality appear to be particle-like in nature.

In other words, the quantum aspect of Einstein's mathematical description concerned only the properties of the phenomenon that were being manifested in the photoelectric effect but did not necessarily capture the nature of what gave quanta their properties under such conditions (e.g., what is the origin of the particular constant value of any given quantum package?). As such, Einstein's quanta were a surface phenomenon that were functionally dependent on an underlying set of hidden variables that gave rise to a phenomenon (the photoelectric effect) whose surface or manifested features could be accurately captured through Einstein's particulate approach to the photoelectric effect.

Finally, applying the same sort of analytical perspective that has been used in conjunction with Planck and Einstein, one also might come to the conclusion that Louis de Broglie could be considered to have been both correct and incorrect when he said there is a wavelength that is associated with every material object or particle that is related to the momentum of that particle by means of Planck's constant. De Broglie is correct because the mathematical relationship he is alluding to in his hypothesis does accurately describe certain facets of the phenomena that arise in conjunction with quantum dynamics, but the wavelength that is associated with a particle is part of the mathematical description of the phenomenon to which what is called a particle gives expression, and, therefore, that wavelength does not necessarily capture the nature of what it is that gives rise to the properties being described by de Broglie through the language of waves.

Einstein objected to the idea of considering statistics and probabilities as being the final word that could be said concerning the nature of reality. He believed in the reality of a set of hidden variables or underlying processes and/or entities and/or forces that could account for why the statistics and probabilities used in quantum descriptions had the properties they did.

There is a similar sort of argument that might be directed toward those who wish to claim that wave-like and/or particle-like descriptions are the final word concerning the nature of the reality to which such descriptions refer. The wave-like and particle-like terms are similar to the statistical and probabilistic terms that are generated through quantum calculations in the sense that both sets of terms could be considered to be functions of a deeper or hidden dimension of reality that gives rise to phenomena that can be described in terms of particles, waves, statistics, and probabilities but, in fact, cannot be reduced to such terms because those terms are restricted to the surface level of manifestation and do not necessarily accurately represent that which makes such surface level phenomena possible.

Einstein believed in the idea of a realm of hidden variables – that is, some sort of fundamental bedrock of reality -- that existed independently of the statistical and probabilistic mathematical language used to describe various phenomenal events that arose through such underlying dynamics. The problem is, however, there could be more than one level of hidden variables and, therefore, when Einstein accounted for the photoelectric effect by treating light in terms of particle-like quanta, then, in his own way, Einstein might have been committing a similar sort of mistake as those (e.g., Bohr, Heisenberg, Born, and others) with whom he disagreed when they tried to limit the nature of reality to statistical and probabilistic descriptions, and, consequently, Einstein's notion of hidden variables notwithstanding, there might be at least one more level of hidden variables that existed beneath the surface phenomena of light and that constrained light to behave as if it were particle-like even as that level of hidden variables is not necessarily particle-like in character but can, under the right circumstances, give rise to phenomena that can be described effectively through mathematical descriptions that treat such phenomena as if they constitute a reality that is particle-like in nature.

| Quantum Queries |

76

There is a story associated with Richard Feynman, one of the architects of the modern Standard Model of quantum theory, that resonates, to a degree, with the foregoing perspective. Various sources have reported that a graduate student came to Feynman one day after the student had gone through a period of study and reflection that nearly drove the student crazy as he tried to understand what the actual nature of reality was that was being parsed by various kinds of mathematical equations, techniques, and so on. After listening to the student talk about his feelings of perplexity and despair concerning such matters, Feynman is reported to have said: "Just do the calculations ... no one understands what is going on."

Planck, Einstein, de Broglie and many others found ways to mathematically treat observed phenomena that enabled them to generate calculations that solved an array of problems. However, in doing so, they came up with solutions that were vulnerable to conflicting interpretations concerning the nature or meaning of the sort of 'reality' to which their calculations were alluding or attempting to describe.

Feynman was a pragmatist. He was in favor of proceeding in ways that led to generating reliable and effective solutions with respect to various physical problems and, as a result, he was prepared, within certain limits, to put aside questions concerning the ultimate nature of reality ... especially if such questions got in the way of being able to discover methods for solving certain kinds of problems.

Einstein was a realist. While he was quite prepared to just do calculations that were rooted in statistical and probabilistic models without necessarily being able to understand what was transpiring beneath the surface phenomena being described by various mathematical techniques, nonetheless, he fervently believed that beyond the calculations there existed some ultimate realm of physical law consisting of hidden variables – at least they were hidden until they could be identified and properly understood -- that were generating phenomenal properties that could be described through the language of statistics and probabilities.

Although the early theory of quantum mechanics had generated many successes, nonetheless, by the early-to-mid 1920s a variety of problems began to emerge that could not be handled adequately by the initial set of methodological and mathematical approximations that had been devised for describing how certain facets of the physical world operated. These problems included the issue of whether light was particle-like or wave-like in nature but also involved difficulties that arose in conjunction with developing an adequate account of how light was absorbed and emitted by atoms under different conditions.

Consequently, many physicists felt that a new approach of some kind was required ... an approach that would be able to build upon what already had been achieved but a perspective that also could help lead exploration in a different direction that might be able to solve some of the problems that were beginning to accumulate. Over a period of several months in 1925, Werner Heisenberg decided to try his hand at making a constructive contribution to the aforementioned crisis of understanding.

He started out with the intention of attempting to find some workaround in relation to the notion of an electron's orbit. Niels Bohr introduced the idea that an electron might move in an orbit around the nucleus of an atom ... somewhat like the way in which planets travel in a fixed orbit around a sun.

Borh's model of the atom was an attempt to improve on the earlier 'plum pudding' model of atoms in which negatively charged electrons were described as being distributed throughout some sort of positively charged, homogenous medium. While Bohr's planetary-like refashioning of the model of the atom initially attracted a lot of interest (especially with the general public), it had at least one fatal flaw.

If electrons traveled in an orbit like planets around the sun, then electrons would exhibit angular acceleration. That is, electrons would be constantly changing their direction of movement as they sped about their orbit at some constant rate.

Generally speaking, electrons emit light during the process of acceleration (that is, changing directions during the process of moving through an orbit). As a result, electrons would lose energy as they circled about the nucleus, and, eventually, this would mean that electrons were likely to spiral inward toward the nucleus – eventually crashing into the nucleus -- and, in the process, destabilize the atom.

Bohr came up with a solution to the foregoing problem by re-conceptualizing the notion of an orbit. Instead of conceiving of orbits as paths that circled around the nucleus (and, therefore, were vulnerable to losing energy through emitted packets of light), Bohr treated orbitals as stationary states that, somehow, permitted electrons to avoid constantly losing energy while occupying that state.

Although electrons could move from one orbital state to another through processes of absorbing or emitting packages of energy of a discrete nature, electrons were otherwise constrained … especially with respect to any sort of movements that might result in a spiral of destruction that ended in the nucleus. However, Bohr's proposed solution was relatively arbitrary because there was no concrete evidence demonstrating what electrons actually did within the context of an atom.

To be sure, the behavior of electrons was consistent with the conditions that Bohr imposed on the hypothetical stationary state orbitals. However, how any of this actually took place – in an ontological sense -- was unknown because no one could directly observe the manner in which an electron occupied its alleged orbital state or how an electron moved from one such 'orbital' state to another.

Heisenberg wanted to eliminate the notion of an orbital precisely because no one could see the actual dynamics of electrons within Bohr's hypothetical stationary state orbital. Consequently, Heisenberg decided to focus on what could be observed – namely, the packages of energy that were absorbed or emitted by atoms.

In order to accomplish his aim, Heisenberg came up with a mathematical way to calculate answers capable of describing some of what was transpiring during the process of an atom's absorbing or emitting energy. Not entirely sure about the viability of his mathematical construction, Heisenberg left his work with his mentor, Max Born, and Born eventually came to realize that Heisenberg had developed a way of solving certain problems concerning the atom that was connected to the mathematics of matrices.

Heisenberg's breakthrough led to what became known as matrix mechanics ... the use of matrices to represent and make calculations concerning certain aspects of the behavior and properties of electrons in transition. However, the development of matrix mechanics was not solely due to Heisenberg's initial breakthrough since the method was further refined through the collective efforts of Heisenberg, Max Born, and Pascual Jordan.

Without understanding the actual dynamics of electrons within an atom any better than Bohr did, Heisenberg had devised a method of calculating answers that was capable of generating usable and reliable results concerning the behavior of electrons within an atom. Heisenberg's mathematical invention didn't clarify the nature of the underlying reality concerning how electrons did what they did within an atom, but, instead, his perspective served to help scientists to be able to run with Bohr's idea of stationary state orbitals and, in the process, solve problems even though no one really knew what was going on as far as the dynamics were concerned with respect to how electrons actually occupied orbitals (if they did) or were able to move about in orbitals – if that is what they did – or how electrons made the transition between orbital states.

During 1925, and carrying over into 1926, Erwin Schrödinger constructed an equation that he hoped would be able to replace the matrix mechanical model. Although the Heisenberg/Born/Jordan method did offer a mathematical means through which to solve certain problems, that model also was very abstract in character because it did not permit one to have any sort of working image of what was taking place in an atom.

The matrix mechanical model was like a black box. Empirical values could be entered into that computational device, and output values based on such input data could be calculated by running them through a set of matrices that were provided by the model, but one had no understanding of the actual dynamics that were taking place while the model was grinding out its solutions.

Schrödinger's differential equation was intended to serve as a means through which to model how de Broglie's wave function -- that supposedly was associated with each particle -- unfolded over time. Consequently, it became known as a wave equation.

However, there were some difficulties surrounding the meaning of Schrödinger's equation. If one understood Schrödinger's equation to be a description of particles as a waveform, then the equation indicated that such a wave should continue to expand in all directions becoming improbably large in the process, and as the purported wave did so, that description becoming increasingly at odds with a variety of empirical data.

There were a few other problems as well that were caught up in the wave equation. For example, when that equation was used in calculations involving several particles, more than three dimensions were needed to solve problems, but no one really knew how to reconcile those extra dimensions with what was believed to be a three-dimensional reality.

| Quantum Queries |

In addition, the equation involved the use of imaginary numbers. How was one supposed to reconcile the imaginary with the real?

As far as trying to get a picture is concerned with respect to the way reality worked, physicists seemed to be no better off with Schrödinger's wave equation than they were with matrix mechanics. That is, trying to figure out how imaginary numbers, expanding wave functions, and more than three dimensions fit together to give a coherent picture of the dynamics involved in quantum interactions was as problematic as it was to try to understand what the abstract character of matrix mechanics meant with respect to the actual, ontological nature of those dynamics.

Max Born – who had worked with Heisenberg and Jordan in the development of matrix mechanics – found a way to re-fashion Schrödinger's wave equation so that it would be able to generate reliable answers, but his idea came with a price. More specifically, contrary to Schrödinger's initial intention to put forth a mathematical model through which one could obtain a concrete image or sense of quantum dynamics, Born's interpretive reconstruction of Schrödinger's equation jettisoned the notion that the equation described the nature and/or behavior of an electron and, instead, contended that the equation (when properly reworked) described a set of probabilities for locating where an electron might show up in a given set of physical circumstances.

In short, Born was treating Schrödinger's equation as a black box just as matrix mechanics gave expression to a mathematical black box. In each instance, one lost contact -- to some extent – with reality in order to be able to establish a method for generating calculations that gave rise to workable solutions for various problems in quantum dynamics.

Schrödinger's equation didn't describe the dynamics of a particle. It constituted a way of descriptively capturing the possible outcomes of those dynamics.

What were electrons doing prior to such calculations? No one knew!

How did electrons give rise to such probabilities? No one knew!

Why should Schrödinger's equation be able to describe where one might find a given particle or why that particle could be associated with such a probability distribution? No one knew!

How and why did electrons pop up in one location rather than another? No one knew!

Did Schrödinger's equation actually describe reality? To some extent it did, and to some extent, it did not.

The equation made precise predictions that usually reflected certain aspects of ensuing reality. However, no one understood what made those predictions turn out correctly in the way they did.

Unfortunately, somewhere along the line, quantum physicists reified methodology and began to treat probability distributions as ontological in character rather than merely being descriptive of certain kinds of outcomes. Niels Bohr was one of the chief architects of this approach to quantum phenomena ... an approach that came to be known as the Copenhagen interpretation of quantum mechanics.

For instance, consider Heisenberg's uncertainty principle. As originally conceived, the uncertainty principle was about how one's ability to simultaneously determine precise values for certain pairs of variables -- such as momentum and position – was methodologically limited.

The more accurately one sought to determine the momentum for a given particle, the more such an attempt would interfere with one's ability to accurately determine the position of that particle. The reverse also was true.

The formal expression of the uncertainty principle is: $\sigma_x \sigma_p \geq \hbar/2$ (developed by Earle Kennard and Hermann Weyl during the period 1927-1928). This inequality states that the standard deviation of position and the standard deviation of momentum, when considered together, are greater than or equal to the reduced form of Planck's constant (that is, $h/2\pi$).

When Heisenberg made the idea of the uncertainty principle public through the pages of a paper on the topic in 1927, he used the example of a microscope. More specifically, if one were to use light to determine the precise position of a particle, the extent to which that position could be accurately resolved would depend on the wavelength of the light being used to measure position.

The smaller the wavelength of light employed, then the more energy will be contained in that light. While such a wavelength can help one to narrow down the measured position of a particle, that same light energy also will impact the momentum of the particle being measured.

The very methodological process of trying to simultaneously measure position and momentum would affect the accuracy with which one could determine values for the conjugate properties of that particle (such as position and momentum). If one attempted to pin down the position of the particle, this attempt to establish greater precision would affect the precise measurement of momentum, and if one tried to determine the momentum of a particle using light energy, then such a measurement would adversely affect the precision with which one could determine the position of that same particle.

When Heisenberg discussed his microscope example with Bohr, Bohr accepted – to a degree -- the uncertainty principle that was being illustrated through the microscope scenario, but Bohr maintained that Heisenberg's microscope example carried unwanted implications because it suggested not only that there was a definite path being followed by a particle that could be impacted by the light energy being used to measure things, but, as well, Heisenberg's microscope example suggested that during instances when nothing was being measured, then, particles would have some determinate ontological position and momentum that was on-going independently of the measuring process.

Bohr believed that Born's probabilistic version of Schrödinger's equation indicated that all one could talk about were the probabilities associated with different possible paths of a particle. One could not talk about the paths themselves.

As a result, Bohr insisted that particles did not move along a specific path. In doing so, he confused and conflated mathematical description with ontology.

Bohr did not possess any evidence indicating that particles could not or did not follow specific paths. All he had were the methods of matrix mechanics and Born's reworking of Schrödinger's equation.

Each of those computational techniques entailed a hermeneutical variation on Heisenberg's uncertainty principle. In other words, the more one focused on determining the character of the behavioral outcomes of quantum processes, the less one understood how such outcomes were possible in the first place.

Bohr was correct when he maintained that being able to calculate a probability for where a particle might appear did not necessarily imply that such a particle had to follow a particular path in order to be able to arrive at the point of manifestation predicted by such a calculation. However, Bohr was wrong to suppose that Born's reworking of Schrödinger's equation necessarily implied that particles did not, and could not, follow a particular path to the point where the particle was predicted as having a certain probability of showing up.

The fact of the matter is, physicists didn't know why matrix mechanics or Schrödinger's equation worked. They just knew that they did.

Rather than admit that Schrödinger and Born had stumbled on to something that worked but did so for unknown reasons, some individuals – such as Bohr – decided to invest the mysteries surrounding the functionality of the wave equation with ontological value. Probability distributions did not just have descriptive value, but, as well, they reflected the nature of reality itself.

Bohr maintained that a particle didn't have momentum, position, or any other quantum property until the act of measurement generated a specific value. Somehow (and Bohr never explained how), the process of measurement transformed a dimension of the probable into a real world value.

Yet, one wonders about the nature of such a transformation process. If particles have no ontological value until a measurement is made, how does a probability get turned into a reality?

In addition, what sort of reality is a probability? A probability is a description of a certain facet of reality so, at the very least, probability refers to certain behavioral properties of that which is being described.

Yet, beyond such a descriptive quality, what reality did probabilities have? What made that to which probabilities referred have the properties being alluded to through probabilities?

What is the nature of the dynamic that allegedly takes place on the quantum level involving the manner in which the process of measurement somehow engages probability to produce an ontological entity with concrete values? How does a measurement generate something real out of – seemingly – a non-ontological entity (i.e., probability)?

Between 1927 and 1930, Einstein attempted, informally, to poke holes in Bohr's belief about the manner in which uncertainty allegedly permeated reality. Bohr maintained that the issue of uncertainty was inherent in the nature of things and was not just a function of being unable to come up with a methodological means for detecting the presence of hidden variables and/or an inability to demonstrate that there might be experimental ways to uncover more information concerning a situation than the uncertainty principle said was possible.

While attending several Solvay Congresses (one of the primary scientific venues through which scientists discussed their ideas and discoveries with one another), Einstein devised a number of thought experiments that were intended to challenge Bohr's aforementioned position concerning uncertainty. On each occasion, Bohr was able to point out difficulties with one or another aspect of Einstein's thought experiments.

However, in repelling Einstein's various challenges, Bohr wasn't proving that his own perspective was correct. Bohr was demonstrating that there were problems with Einstein's way of engaging the uncertainty issue.

Einstein was trying to show that reality was not inherently uncertain in nature. However, even if Einstein had only been trying to demonstrate that uncertainty was a purely methodological issue and not an ontological one, Bohr's counterarguments would have succeeded because there were, in fact, conceptual problems with each of Einstein's thought experiments ... details and considerations involving the issue of uncertainty that Einstein had not taken into account when devising his theoretical challenges to Bohr's position.

For instance, consider Einstein's initial challenge to Borh's perspective. Einstein described an experiment in which a beam of electrons is fired toward a narrow slit, and after passing through the slit, the electrons begin to behave in a wave-like fashion and fan out instead of acting as if they were particle-like in nature and, thus, continuing to move in a straight line.

Einstein's thought experiment specifies that on the other side of the slit area, there is, at some arbitrary distance from that slit, a semicircular shaped film. According to the Copenhagen interpretation of quantum theory, one cannot know where the electron is or in what state it is in until it strikes the film at some point, thereby blackening the film and, in the process, generating a measurement event.

Einstein found such a description troubling. More specifically, if all quantum theory could do was to compute a probability distribution that described where the electron might engage the film, then Einstein argued that this was tantamount to saying that every portion of film had some random probability that could be assigned to it and which would indicate the likelihood of that portion of the screen being impacted by the electron.

However, if the Copenhagen interpretation of quantum theory were correct, then Einstein believed that the foregoing sort of description would seem to mean that every other portion of the film that was not impacted by the electron when the latter entity engaged the film would – in a mysterious fashion -- have to be able to receive communication of some kind informing it not to turn black. According to Einstein, such communication would have to be instantaneous – meaning that all points in the semicircular film were connected in some way.

Bohr's response to Einstein's thought experiment was one of confusion. He didn't seem to grasp the thrust of Einstein's argument.

If someone claims confusion, trying to figure out the nature of the confusion becomes a rather difficult – and, perhaps, somewhat fruitless --exercise ... at least as far as Bohr is concerned. Nevertheless, there is a certain amount of confusion inherent in Einstein's thought experiment, and bringing the nature of that confusion to light could be a productive exercise.

In Einstein's thought experiment, once an electron is shot in the direction of the narrow slit arrow, no one knows what is going on. This is not necessarily because there is some kind of uncertainty that is inherent in the nature of reality but because there are no empirical eyes on the electron prior to the time it shows up at the slit ... and, perhaps, not even then.

At the slit, Einstein indicates that the electron appears to undergo a transition due to diffraction and, as a result, the electron begins to spread out as it moves toward the film. Diffraction is inferred to have occurred because of subsequent results that manifest themselves in conjunction with the film, but no one has actually seen what happens when the electron passes through the slit, and, therefore, the precise character of quantum dynamics taking place at the slit are unknown.

At one point, the electron appears to behave in a particle-like fashion (as it is sent toward the slit area). At another point, the electron seems to behave in a wave-like fashion (apparently due to the process of diffraction as the electron engages the narrow slit area).

Quantum theory predicts how an electron behaves in a certain set of circumstances (such as Einstein's thought experiment). That theory can't explain how an electron is able to manifest particle-like properties at some points while manifesting wave-like properties at other points.

Quantum theory computes a probability distribution to describe what will happen once a beam of electrons is sent toward the slit. No one knows what the electron is doing (or not doing) once it has been released, and no one knows what the precise nature of the dynamic is that takes place at the slit area, and no knows where the electron will engage the film, but the computed probability distribution descriptively encompasses all possibilities involving such an impact event.

Probabilities do not necessarily determine where an electron will blacken the film. Rather, electrons act in a way that gives expression to some aspect of the probability distribution used to describe that behavior.

The probability distribution profile is useful because it accurately describes how electrons behave under a given set of circumstances. Nonetheless, the probability distribution gives expression to only one kind of understanding ... namely, the likelihood that an electron will blacken the film at one point rather than another.

Nothing else about the behavior or state of the electron is known from the time the electron is released until to the time when a point on the film becomes darkened. Everything that is said -- about the behavior or state of the electron between: The moment of release, until the time when the film is reached -- is an inference based on observed behaviors (the darkening of the film).

The probability distribution can't tell anyone why electrons behave in the manner described by such a distribution. The distribution only indicates that this is the way things seem to turn out.

Statisticians and probability theorists are very good at coming up with figures to assign probabilities for what proportion of a given population is likely to commit suicide or develop cancer or die in a traffic accident or get divorced over the course of, say, a year. However, those statisticians cannot tell one when, why, where, or how any of the foregoing possibilities will happen ... only that such events are likely to take place in the way that is indicated by various probability functions.

To a fair degree, the foregoing considerations also hold in relation to quantum mechanics. Physicists can generate computations concerning the probability that a certain kind of quantum event will take place, but they can't always say when, why, where, or how such events will take place.

Probability distributions don't cause suicide, cancer, traffic accidents, or divorce. Similarly, probability distributions don't cause quantum events, but as is true in conjunction with suicide and so on, what goes on in quantum dynamics does tend to reflect the probability distributions that have been drawn up to describe certain facets of those dynamics.

Einstein devised his foregoing thought-experiment because he thought that the Copenhagen interpretation of quantum mechanics committed its proponents to maintaining that quantum events were random occurrences that could manifest themselves on any portion of the film and, consequently, the probability distribution appeared to be playing some sort of a causal role with respect to where electrons actually showed up ... after all, if there were nothing beyond the probability distributions (i.e., no hidden variables determining why one thing rather than another took place), then, what else could account for why an electron blackened one part of the film rather than a different point on the film -- except uncertain, random events.

As a result, Einstein was concerned that if quantum events were really a random phenomenon and if probability distributions were somehow playing a causal role in determining where the film would become blackened, then, presumably, all of the other points on the screen would have to be informed – in some instantaneous fashion -- not to become blackened.

What is really meant by the idea that something is random? This issue was discussed, to some extent, in Chapter Two of Final Jeopardy: The Reality Problem, Volume I.

Whenever one uses the term "random", one is either stating a position of ignorance concerning why a given dynamic occurs in the way it does or one is putting forth an ontological theory about how the universe operates on its most fundamental level. Bohr was doing both in his Copenhagen theory of interpretation concerning quantum events.

Bohr had turned the uncertainty principle from a methodological problem (the limits of precision) into an ontological statement about the nature of reality – and reportedly reduced Heisenberg to tears while doing so. According to Bohr, reality was inherently uncertain and random in character.

However, statements involving probability distributions had an ambiguity about them that helps to lend confusion to the Einstein-Bohr dust-up. Such statements could be used in conjunction with mathematical computations that generated descriptions about certain aspects of quantum behavior, or they could be used to give expression to a theory concerning the inherent nature of reality ... that it is uncertain, random, and probabilistic in nature.

There was no evidence to demonstrate that reality could be reduced to being nothing more than uncertain, random evens as the Copenhagen theory of interpretation claimed. However, the precision and heuristic properties of probabilistic descriptions were leveraged to make unfounded ontological claims concerning the nature of reality.

If reality were merely a function of random, probabilistic events, then, how – Einstein wanted to know -- did any given point on the film in the thought experiment "know" when, or when not, to become blackened? How were the points on the film informed about what should happen?

How – Einstein wanted to know -- could one suppose that such communication would be able to take place in any other manner than an instantaneous one? In other words, how did all points on the film become informed at the same time about which point would become blackened?

Einstein was trying to visualize how things worked. He wanted to understand the nature of the dynamic that was taking place when the electron engaged the film, and a probability distribution couldn't provide him with the sort of explanation he was seeking.

Einstein didn't have a problem with the idea that probability distributions could be used to describe the behavior of events, and he had used those sorts of tools in his own work. Nonetheless, he did have a problem with the idea that the Copenhagen school of interpretation was using probability distributions as the basis for an ontological account of why quantum events occurred in the way they did.

On the one hand, Bohr was enlisting probability distributions as a means of giving expression to his ontological theory concerning the random, uncertain, nature of reality. On the other hand, Einstein was trying to figure out how one went from the descriptive, methodological properties of probability distributions to an ontological understanding of those same probability distributions.

Einstein was asking a very straightforward question: What is the nature of the dynamic that is taking place at the surface of the film? How does one point within a given probability distribution become or transition into a particular reality at a certain point on the film?

From Bohr's perspective, such questions were ignoring the "fact" (according to Bohr) that reality was inherently uncertain and random, and, therefore, Einstein's question couldn't be answered. Things happened and, as well, certain probabilistic descriptions of what took place were possible, but that was the end of the matter.

For Bohr, probability distributions could describe – within limits – what might happen when an electron was sent toward a narrow slit with a semicircular film on the far side. However, those probability distributions -- when considered through the filters of the Heisenberg uncertainty principle (at least as interpreted by Bohr) – went as far as one could take things ontologically because such probability distributions supposedly gave expression to the fundamental nature of reality … there was nothing of a hidden nature that was happening beneath those random, uncertain, probabilistic events

So, as far as Bohr was concerned, to ask how the film "knew" which point would become blackened and how such information would be communicated to all of the other points on the film was to introduce confusion into the matter. The point on the film that became blackened was a random event, and, consequently, there was no need for anything to be communicated to the other points on the film because random events are self-selecting as far as where they show up is concerned.

Bohr was able to fend off Einstein's attacks because, in one way or another, he could point out problems with Einstein's thought experiments. Bohr never had to prove that his own ontological position was true, but, rather, he only had to point out the shortcomings of Einstein's challenges.

Perhaps instead of attacking Bohr's theoretical position, Einstein should have demanded that Bohr defend his own position and prove that the inherent nature of reality was uncertain and random. If Einstein had changed his tactics to something along the foregoing lines, Bohr very likely would have floundered as badly as Einstein's thought experiments did since it is often easier to point out the flaws in someone else's way of trying to account for observed phenomena than it is to put forth a viable argument of one's own concerning the nature of reality.

Bohr didn't actually demonstrate what the nature of reality is. He pretty much evaded that issue altogether.

Using probability distributions as a methodological technique through which to describe certain facets of reality can be defended because those techniques are capable of producing reliable, heuristically valuable descriptions. Nevertheless, trying to prove that such descriptions also account for the inherent nature of reality is a very different matter ... a proof that Bohr never successfully accomplished – rather, he merely assumed that his position was correct and proceeded from there.

Bohr's interests were assisted by the fact that the methodological/descriptive side of his hermeneutical position worked. Einstein couldn't argue with scientific success.

Indeed, since Einstein's thought experiments were unable to penetrate Bohr's theoretical defenses concerning the issues of uncertainty and randomness, Einstein was made to appear as if he were arguing for the wrong side of the issue. Therefore, many individuals seemed to miss the fact that Bohr never demonstrated that reality was inherently uncertain and random but, rather, Bohr had shown only that the particular thought experiments brought forth by Einstein were flawed in one way or another.

Like a magician, Bohr made use of a form of misdirection during his point-counterpoint with Einstein. That is, Bohr kept everyone's focus – including Einstein's -- on the problems with the thought experiments being devised by Einstein.

This tactic led many people to presume that Bohr had won the debate concerning the nature of reality when, in truth, the debate between Bohr and Einstein was never about the tenability of Bohr's beliefs concerning the nature of reality (although Einstein tried his best to push the discussion in that direction). Instead, the debates that occurred between Einstein and Bohr during several of the Solvay Congresses always tended to gravitate back toward the problems that afflicted Einstein's thought experiments and, therefore, discussion moved away from the rather glaring fact that Bohr really didn't have any proof that his Copenhagen theory of interpretation was a correct account of, or explanation for, the nature of reality.

To be sure, the Copenhagen interpretation of quantum mechanics was – within certain limits -- a coherent, meaningful description of reality, and it was internally consistent. Furthermore, the probability distributions that arose in conjunction with **Schrödinger's equation** were compatible with – but not necessarily proof of, or evidence for -- the idea that reality was inherently uncertain, random and probabilistic.

Nonetheless, at no point did Bohr do anything to show that such probabilistic descriptions were anything more than methodological tools that yielded useful solutions in relation to certain physical problems. Bohr interpreted those descriptions in an ontological manner, but he never demonstrated that he was justified in doing so.

Einstein's thought experiments might not have proven that hidden variables existed. However, Bohr never demonstrated that such hidden variables didn't exist … only that there were flaws with Einstein's attempt to find fault with Bohr's interpretive perspective concerning the nature of reality.

Einstein took one more kick at the can bearing the label of the Copenhagen theory of quantum mechanics. The kick did not assume the form of an informal point-counterpoint with Bohr during one, or another, of the Solvay Congresses, but, instead, the kick assumed a more formal expression in the form of a May 15, 1935 paper entitled: "*Can Quantum Mechanical Descriptions of Physical Reality Be Considered Complete*" that was published in the *Physical Review* and authored by Einstein, Boris Podolsky and Nathan Rosen.

The paper placed the Einstein-Bohr debate in an either-or context. Either the Copenhagen theory of quantum interpretation was wrong in denying the existence of hidden variables, or the principle of locality (which is at the heart of Einstein's realism) was false and, as a result, two objects that are separated by distance can, in fact, influence one another without anything being communicated across the distance that separates the two objects.

Of course, Einstein, Podolsky, and Rosen were firmly of the opinion that the principle of locality was inviolable. Their paper was an attempt to point out – or, at least, allude to – the existence of problems with the position of those – such as Bohr -- who believed that nature was inherently uncertain, random, and probabilistic and, therefore, who believed that, among other things, the principle of locality did not necessarily hold.

The EPR paper puts forth an argument that is a variation on Einstein's previously discussed initial challenge to Bohr's theoretical position concerning the nature of reality. To briefly summarize: The latter argument involved questions about how the different parts of a semicircular film "knew" which point on the film should turn black and which points should not turn black if probability distributions somehow led to an electron ending up at one point rather than another on the surface of the semicircular film that was set up, hypothetically, beyond the slit through which an electron traveled and that, supposedly, led to the diffraction of the electron and an ensuing spreading out of a wave of some kind that moved toward the semicircular film.

In the 1935 EPR thought experiment, Einstein, Podolsky and Rosen imagine that a particle of some kind decays and in the process yields two further particles that speed off in opposite directions from one another. According to the Copenhagen theory of quantum interpretation, there are no set quantum properties that exist in either of the foregoing particles, but, instead, only probabilities involving possible values can be associated with such particles, and when a measurement is made, we come to know which of the predicted probabilities are manifested into existence.

The two particles supposedly are related in such a way that when information becomes known about one of the particles, then, immediately, something about the other particle is also known. For example, if one makes a measurement involving momentum with respect to either of the two particles, then one immediately knows the momentum of the other, unmeasured particle.

However, as far as EPR are concerned, this is precisely the problem. If – according to the Copenhagen school of thought – quantum properties don't exist until a measurement is made, then how does the particle that is unmeasured nevertheless instantaneously "know" what momentum to manifest given that the two particle are separated by a distance across which there has been no communication – as far as anyone knows -- concerning the issue of momentum.

| Quantum Queries |

EPR state that given the distance that separates the two particles, there is no way that one particle (the unmeasured one) could know what has transpired with the other particle (the measured one). Consequently, by process of elimination, then, seemingly, the only possibility that plausibly can account for the hypothetical situation being described by EPR is to conclude that the unmeasured particle always had a fixed momentum that is independent of the process of measurement.

The EPR paper continues on to make the same sort of point with respect to the issue of position. After doing so, the thrust of the EPR argument is that one is faced with having to choose between, on the one hand, the incompleteness of quantum theory or, on the other hand, the idea that the principle of locality has been violated in some way, and since there no evidence to indicate that the principle of locality has been violated, then it has been demonstrated that quantum properties are real and independent of measurement, and, therefore, quantum theory is incomplete because it can't account for the existence of such on-going, real, quantum properties.

Of course, in principle, there is nothing to prevent Bohr, or anyone else, from arguing that, in point of fact, we don't know what the momentum or position of the unmeasured particle is because it hasn't been measured. Such an argument might maintain that EPR are assuming that the two particles will have identical momentums but, perhaps, there is some small probability that the two particles will have different momentums under certain conditions, but this can't be determined until the second particle is measured.

If and when the second particle is measured for the property of momentum and that property turns out to be identical with the value for the previously measured particle, then, this is only because – or, so, it might be argued -- the probability distribution gives expression to that value at the time of measurement and not because the particles have fixed quantum properties existing between measurements. The foregoing argument is somewhat tautological in character – and a touch sophomoric -- but it could be a possible response to the EPR paper.

However, a different approach subsequently was taken by a number of experimental physicists. The result of those experiments tended to bring the principle of locality into question rather than to demonstrate that quantum theory is incomplete.

Before a claim can be experimentally probed to determine whether, or not, that claim might be true or false, one needs to be able to state things in a manner that could be empirically tractable or amenable to testing. John Bell, an Irish physicist, was the individual who helped clarify some of the issues in this regard.

Instead of focusing on the momentum or position of a particle, Bell decided to make spin the quantum property that is to be measured. The nature of spin is somewhat obscure.

Spin comes in two varieties – up and down. However, the issue of spin is not a straightforward matter, and, consequently, there is a lack of clarity concerning what actually is going on when the property of spin is mentioned.

The idea of spin arose in a context that described electrons moving about the nucleus much as a planet moves about a sun (Bohr's idea). According to such a model, electrons operated within the parameters of an orbital, and spin was likened to the manner in which planets spin on their own axis while also traveling about the sun.

The problem with the foregoing image is that – as previously indicated – one runs into problems by supposing that electrons travel about a nucleus as planets move about a sun because under those conditions electrons would lose energy during such a process of orbiting and, as a result, those particles eventually will spiral into the nucleus. As noted earlier, Bohr solved the foregoing idea by re-conceptualizing the notion of an orbit and transforming it into some sort of stationary state of fixed energy ... although Bohr was never very clear about what any of this meant – or how it was possible -- other than to indicate that doing so permitted one to avoid the 'electron-spiraling-into-the-nucleus' problem.

Electrons occupied orbitals. Nonetheless, what, if anything, electrons did in those orbitals was unknown.

Similarly, electrons possessed the property of spin, and experimental results indicated, as previously noted, that that property came in two varieties that arbitrarily were dubbed: "up" and "down". However, just as no one knew what went on within such orbitals, no one really knew what spin actually entailed.

Some individuals referred to spin as being a kind of angular momentum associated with an electron while it did whatever it did within an orbit. However, since no one really knew what electrons were doing while in "orbit" – the orbit-term is potentially misleading because no one knows whether, or not, an electron actually circles or orbits about a nucleus – and, therefore, one can't be sure that the electron spins about an axis while in orbit.

The energy state of an electron can be measured, and, as well, the spin state of an electron can be determined – that is, one can determine whether the state of the electron is 'up' or 'down'. Yet, none of this information can tell one what it means for an electron to have spin while occupying a given orbit.

The notion of spin is useful because it permits one to be able to distinguish between particles that possess that property even though we don't quite know what spin actually is. In fact, the capacity of certain particles to exhibit different conditions of spin (I.e., 'up' or 'down') plays a crucial role in experiments involving the issue of entanglement.

Bell imagined a variant of the EPR thought experiment in which a particle gives rise to two further particles that take off in opposite directions. At some distance in both directions, a detector is set up to capture the state of spin of the particles.

However, the detectors are not set up in the same way. They are calibrated for different angles to determine whether the state of the particle was up or down at those two angles.

If the detectors are aligned with one another, Bell believed that one could come up with a hidden variable theory that was capable of accounting for what is observed at the two detectors. But, if the detectors were not aligned with one another – that is, if they were set at different angles – then, a hidden variable theory might have difficulty accounting for certain kinds of results.

More specifically, Bell gave a quantitative face to his thought experiment. If one measured the spin of the two particles over a range of angles that were independent of one another (i.e., not aligned), a set of conditions would have been established through which one could differentiate between what a quantum theory would predict and what a hidden variable theory would predict that operated in accordance with the principle of locality.

According to Bell, if a series of experiments were run and if the statistical treatment of the collective results from those experiments fell outside a given set of parameters, then, the principle of locality would have been demonstrated to be incorrect. The statistical parameters came to be known as Bell's inequality or Bell's theorem.

Bell was not trying to prove that either Einstein's notion of locality was correct or that quantum theory was correct. Instead, he was trying to devise a way that would be able to empirically distinguish between the two perspectives, and Bell's theorem -- Bells' inequality -- gave expression to a prediction that, under the right set of circumstances, could be tested with respect to those differences.

Since Bell developed his inequality or theorem, a number of experimentalists have tested his prediction. For example, John Clauser and Stuart Freedman did so in 1972, as did Alain Aspect, along with several other individuals, in 1981.

While different experimental techniques were used in the foregoing empirical tests, the results all seemed to show that – statistically speaking – the modes of detection employed in the respective experiments indicated that the separated particles seemed to be sensitive to their respective conditions in ways that could not be explained in terms of hidden variable theories or that could not be understood if the principle of locality were in force. Since the running of the aforementioned experiments, various criticisms have arisen as to whether, or not, the different experiments were as loophole free as would be needed to definitively demonstrate that Einstein was wrong – as far as hidden variable and the principle of locality were concerned – or that quantum theory was correct.

The foregoing criticisms tend to revolve around issues of experimental design (such as whether, or not, the detectors employed in the experiment are engaged in a process of fair sampling, or, alternatively, whether, or not, there might have been some way in which particles were transmitting their quantum states to one another without necessarily violating the principle of locality). In addition, some of the criticisms explored the possibility that the aforementioned experimental results still might be capable of being explained through some kind of hidden variable theory.

In what follows, I am going to accept as given – with certain qualifications to be noted subsequently -- that the phenomenon of entanglement has been experimentally demonstrated. In other words, I am accepting the idea that experiments have shown that Bell's inequality has been exceeded in ways that cannot be accounted for in a plausible fashion when approached from traditional notions of locality which require that in order for two objects to affect one another, there must be some form of spatial and temporal connection between those two objects that permits a communication of information across the distance that separates them in a manner that does not entail violating the speed of light.

To date, no one knows what the phenomenon of entanglement actually involves. Like spin, the phenomenon of entanglement can be detected, but what, exactly, is being detected is unknown.

Have any of the experiments that were alluded to earlier demonstrated that entanglement is a purely random, uncertain, probabilistic phenomenon? Or, have those experiments only shown that such experimental results are consistent with the capacity of quantum theory to be able to accurately predict certain kinds of outcomes?

No one disputes the ability of quantum theory to be very good at what it does – namely, describing and predicting the behavioral dynamics of various kinds of particles under different physical circumstances. On the other hand, what has been at issue -- almost from the very beginning -- are the following issues: What exactly is the nature of *that* which quantum dynamics does such a good job of describing ... at least in terms of some of its behavioral feature? What, if anything, makes the phenomena possible that are being so precisely described?

Bells' inequality states that if one collects enough data in relation to what occurs when two detectors have their angles set in a non-aligned fashion with respect to two particles that have separated themselves from one another, then, statistically speaking, sufficient evidence can be compiled that will enable one to be able to determine whether quantum theory or Einstein's realism better reflect that data. According to a number of experiments, quantum theory is better able to handle predicting the outcomes of such experiments than is any known or presently conceivable hidden variable theory that observes the principle of locality.

Nonetheless, despite the foregoing results, one continues to be confronted with the same question that bothered Einstein. How do probability distributions get turned into the real events that are statistically compiled over the course of a series of experiments? What is the nature of the transition that takes one from a theoretical prediction to an actual event if there are no concrete quantum properties present in the particles that are about to be measured?

When allegedly random, uncertain probabilities engage the process of measurement, something happens? Isn't this process of engagement – whatever the specific character of its dynamics might be -- an expression of the principle of locality at work?

Perhaps, such an expression of locality – if that is what it is -- does not operate in quite the way that Einstein envisioned. Nevertheless, something is affecting particles with no inherent, fixed quantum properties at one point and inducing those blank slates to begin exhibiting specific quantum values at some subsequent point, and isn't the issue of causality (being able to affect or induce dynamics) at the heart of the principle of locality?

Does the measurement process cause a particle – which, supposedly, possesses no fixed quantum properties -- to suddenly acquire certain, concrete values, and, if so, how does this take place? Or, do random, uncertain, probabilities cause a measurement to take on certain, concrete values, and, if so, how does this take place?

The entanglement experiments might indicate that Einstein's notion of, or approach to, the issue of locality is problematic. However, in the light of the foregoing comments, those experiments have not necessarily completely eliminated the issue of locality from consideration.

Furthermore, just because one is willing to cede the fact that the phenomenon of entanglement doesn't seem to involve the principle of locality when that idea is construed in terms of signals being communicated across the spatial distance separating two particles, this does not necessarily force one to abandon the principle of locality. Maybe there are non-spatial dimensions involved, and, maybe, such non-spatial dimensions contiguously link particles in ways that fall outside the normal way of thinking about reality as a function of relationships that take place within three spatial dimensions or even x-spatial dimensions.

Time is a possible example of a non-spatial dimension. Unfortunately, the nature of time appears to have become hermeneutically distorted through its spatialization by scientists so that (at least in behavioral terms) the idea of time becomes treatable via mathematically-based descriptions and representations of reality that often carry a geometric bias that requires dimensions to be spatial in character (at least descriptively speaking) rather than being non-spatial and, possibly, resistant to being reduced to any sort of metric that can be spatially represented. (There will be more on this issue of dimensionality in the chapter on mathematics that will appear in a later volume of this book.)

Does time occupy space, or does space occupy time? Or, perhaps, neither of the foregoing two ways of hermeneutically engaging reality is correct.

Maybe space is just one kind of dimension, consisting of at least three degrees of freedom. Perhaps there could be other kinds of non-spatial dimensions – other than time – that might be operative and capable of causally linking two particles that do not exhibit the sort of locality that is said to exist with respect to phenomena that are restricted to just the dimensions of space and time.

If one doesn't know what entanglement is, how can one make any definitive statement about what it does, or does not, involve? If one doesn't know what entanglement is or what makes it possible, then, from a certain perspective, one cannot automatically rule out the possibility that entanglement is itself a function of hidden variables … although, in some respects, such hidden variables – if they are present -- might operate

| Quantum Queries |

quite differently than in the way that Einstein originally felt that the physical realm might work.

The fact that no one has been able to figure out what such a hidden variable theory might look like does not constitute evidence that a hidden variable theory does not, or cannot, exist. Just as the LHC experiments that yielded evidence for the existence of the Higgs field in 2011 eliminated some versions of Supersymmetry from consideration but did not eliminate all forms of Supersymmetry as possible explanations for what was observed in those experiments, so too, the entanglement experiments have eliminated some versions of hidden variables from consideration but those experiments have not necessarily shut the door on the possibility that some other version of hidden variable theory might be able to successful account for the results of entanglement experiments.

One cannot use ignorance about the nature of reality as proof of anything. Indeed, if one is going to use one's inability to imagine how to solve a problem in a given way at a given time (e.g., through a hidden variable theory) as a legitimate reason for concluding that such a current inability must mean no such solution exists, then, the pursuit of understanding concerning the nature of reality would have come to a screeching halt more than a thousand years ago.

The entanglement experiments have not demonstrated that there are no hidden variables present in such experiments. Rather, those experiments have shown that if hidden variables are operating beneath the surface of the observed results, then, those variables will have to have a set of properties that is different from anything that is currently known or understood about that kind of theory.

Another thing that the entanglement experiments have not done is to demonstrate that the Copenhagen interpretation of quantum mechanics is correct. Instead, what has been shown by those experiments is that quantum methodology does a much better job of predicting the outcome of those experiments than does any known theory of hidden variables.

By restricting itself to predicting behavioral properties of particles and, for the most, simultaneously jettisoning any considerations about what makes such behavior possible, quantum methodology has managed to accumulate a variety of algorithms that permit scientists to be able to accurately predict the likely behavioral properties of particles under different conditions. However, unless one is totally spellbound by a Pythagorean-like obsession with the idea that reality is a function of mathematical entities and that, therefore, mathematical expressions necessarily reflect the essence of reality, one can't help but notice that, thus far, there is a complete absence of any plausible account in quantum theory with respect to how reality came to have the capacity to manifest the properties that are being described through the application of quantum methods, and, as well, there is a complete absence of any plausible account in quantum theory with respect to being able to explain how probability distributions are turned into real events.

The entanglement experiments have not shown that reality is inherently, random, uncertain, and probabilistic ... in other words that the Copenhagen theory of quantum interpretation concerning the nature of reality is correct. Those experiments have only demonstrated that quantum methodology does a better job of predicting the outcomes of such experiments than does any other kind of theory currently known.

Now, someone might wish to argue that if something looks like a duck, walks like a duck, and sounds like a duck, then, what one is seeing must be a duck. In other words, if quantum descriptions make reality look, walk, and sound like something that is random, uncertain, and probabilistic, then, surely, reality must be random, uncertain, and probabilistic.

On the other hand, one also could argue that when one views reality through a certain kind of filter, then, one should not be surprised if what is being viewed through that kind of filter appears to assume some of the properties that are inherent in the nature of such a filter but are not necessarily a reflection of the nature of reality. Quantum methodology constitutes a set of filters through which to engage reality, and, therefore, one cannot necessarily conclude that if such a set of filters makes reality look, walk, and sound like a random, uncertain, and probabilistic phenomenon, then this is an accurate reflection of the character of reality.

Entanglement experiments have not shown that reality looks, walks, and sounds like quantum theory. Rather, those experiments have shown that quantum methodology constitutes a means of filtering certain kinds of experimental information in a manner that is capable of yielding accurate results involving the prediction of certain kinds of particle behavior during those experiments.

When a tox screen is run, that procedure can only detect the features that the procedure is set up to detect. Whatever toxins fall outside the parameters of the capacity of that kind of pre-established methodology to detect or capture will appear to be non-existent, but one cannot, then, subsequently declare, that whatever toxins are missed through the application of that methodology do not exist or that reality is necessarily a function of the way that the tox screen filters things.

One must distinguish between the process of filtering information concerning reality and the nature of reality itself. One cannot automatically suppose that any given filter provides a fully complete or accurate representation of that which is being filtered.

Quantum methodology consists of a set of interrelated filters. Those filters are pre-set to engage reality through random, uncertain, probabilistic descriptions.

Does this mean that reality is random, uncertain, and probabilistic? Not necessarily.

Quantum theory looks for, and organizes data according to, the properties inherent in its methodological protocols. Experiments are set up to generate data that can be run through the filters of those protocols and to be interpreted in terms of how those filters engage the experimental set-up.

In summary, the entanglement experiments – as ingenious as they are – have only established one fact. Under certain conditions, particles appear to have a strange form of behavioral interconnectedness that can be accurately predicted – to a degree -- through quantum methods, while, at the same time, the observed form of behavioral interconnectedness seems to fall beyond the capacity of current theories involving hidden variable or the principle of locality, as understood by Einstein, to be able to explain.

Nonetheless, a whole host of other issues remain in play. For example, and as previously indicated, those experiments have not eliminated all notions of locality from consideration ... only certain versions of locality, such as the one championed by Einstein.

Moreover, the entanglement experiments have not established what entanglement is. All that they have demonstrated is that the phenomenon is real.

In addition, the entanglement experiments have not shown that the nature of reality is inherently random, uncertain, and probabilistic. Instead, those experiments have shown that such methodological protocols are capable of leading to accurate predictions concerning the behavior of particles under certain conditions without necessarily demonstrating that no hidden variable theory would ever be capable of successfully accounting for the results of those experiments.

One of the truly amazing things about quantum theory is that a tremendous quantity of precision is so intimately woven together with an incredible amount of ignorance. For example, at the present time, no one knows what the structure of an electron is, and no one knows how an electron occupies an "orbit", and no one knows what spin is, and no one knows what entanglement is, and no one can tell – in the vernacular of the *Invasion of the Body Snatchers* – how the allegedly blank quantum pod of a particle takes on specific, concrete quantum properties, and, yet, certain aspects of the behavior of the foregoing sorts of unknowns can, nevertheless, be determined with considerable accuracy.

Chapter 4: Massive Problems

Until 1964, physicists hadn't come up with any tenable ideas about how particles acquired mass. This condition of understanding, or lack thereof, had a variety of problematic facets.

For example, the rest mass of a stationary electron has been experimentally determined to be $9.10938291 \times 10^{-31}$ kilograms. The foregoing quantity is a high-order estimate since one cannot – at least to date -- capture an electron in a stationary condition and, therefore, the foregoing quantity is derived from analyzing the mass of moving electrons under various conditions and, then, calculating the nature of the unknown on the basis of what is known.

Notwithstanding the foregoing considerations, the resting mass of an electron is one of the fundamental constants of physics. However, no one understood how electrons came to have such a rest mass.

Yet, if electrons did not have a mass, then, they would travel too quickly to be capable of being captured by the protons and neutrons of an atom. As a result, the universe would have become a very different place than the one we encounter now.

Furthermore, if electrons gave expression to a slightly different resting mass than the one that exists, this also would affect the universe in critical ways. Consequently, a number of questions arise – such as: Why does the resting mass of an electron exhibit the constant value it does, and, can one derive the rest mass of electrons from first principles instead of only being able to plug that value in by hand with respect to this or that equation whenever the rest mass of an electron is needed to make some calculation or other?

In 1964, Peter Higgs, a theoretical physicist from Britain, forged a possible answer to questions concerning the origins of mass by coming up with the idea that there might be a certain kind of field that is capable of conferring mass on particles as those particles moved through such a field. Higgs proceeded to formalize his idea in the language of mathematics.

Around the same time that Higgs's was exploring the foregoing sort of possibility, there were a number of other theorists who envisioned possibilities that were similar to his. However, for a variety of reasons – some arbitrary and some not -- the terms "Higgs mechanism" and "Higgs boson" have been almost universally adopted by scientists to refer to the underlying dynamics of the physical field to which Higgs and others were alluding back in the mid-1960s.

Higgs, himself, referred to the process responsible for conferring mass as the: ABEGHHK'tH mechanism. Each of the letters in the foregoing series refers to an individual (Anderson, Brout, Englert, Guralnik, Hagen, Higgs, Kiibble, and 't Hooft) who played a role in helping to develop and/or advance the idea of a mass-conferring field.

Higgs believed that prior to the Big Bang, particles existed in a massless condition. Yet, within an extremely small flicker of time following the beginning of the universe, a spontaneous form of symmetry breaking was hypothesized to have taken place that led to the interaction of particles with a previously dormant Higgs field and, to some extent, mass was born.

The "small flicker of time" required for an allegedly dormant Higgs field to become active has been estimated by some scientists to have lasted just one picosecond – that is, a trillionth of a second -- following the advent of the Big Bang. What caused the Higgs field to turn on is not known.

Some scientists have hypothesized that the Higgs field switched on when the temperature of the universe cooled to a level that was conducive to the Higgs field transitioning from a dormant state to an active one. However, while it is theoretically convenient to make the switching on of the Higgs field to be temperature dependent, why this should be so is not necessarily immediately clear.

In other words, one would like to know how a drop in temperature turns the Higgs field on – if this is what took place. Resolving that issue is not necessarily obvious or straightforward.

Conceivably, the switching on of a dormant Higgs field could have been a function of some kind of spontaneous symmetry breaking process that was unrelated to temperature even though there might have been some sort of non-causal correlation between the two (i.e., spontaneous symmetry breaking and temperature). However, if there were such a process of non-temperature dependent spontaneous symmetry breaking that occurred prior to, or during, the Big Bang, the nature of that process is currently unknown.

Higgs wasn't responsible for the idea of spontaneous symmetry breaking. Yoichiro Nambu had introduced that idea earlier in 1960.

Spontaneous symmetry breaking occurs when a system transitions – through some means or mechanism -- from a state of symmetry to a condition of asymmetry. For instance, suppose there is a circular table that is set up in a symmetrical fashion with dishes, silverware, napkins and glasses of water – a symmetry that is ruptured when one of the diners sitting at that table first selects a drinking glass from a place on the table that is to his or her right or left and, thereby, imposes an asymmetrical pattern on the other diners at the table with respect to which glass – to the left or the right -- can be selected in an ordered manner so that everyone seated at the table has his or her own glass from which to drink.

Yoichiro Nambu had written a paper that explored how a form of spontaneous symmetry breaking might have been able to give rise to mass in fundamental particles. His idea was inspired by the manner in which superconductors had been discovered to work during the latter part of the previous decade.

Normal conductors – such as metals – have an internal dynamic that permits electrons to become organized in lattices that establish many degrees of freedom through which negative charges might flow in and about the positive ions that make up a given conducting medium. This flow of electrons in conductors is temperature sensitive.

Thus, when a conductor is heated, it tends to lose some of its capacity to conduct electrical currents. This is because as the temperature of the material starts to rise, its latticework of electrons begins to exhibit perturbations or vibrations with rising temperatures, and this impedes the free flow of electrons.

On the other hand, when the temperature of a conductor is lowered, perturbations in the lattice tend to lessen. As a result, there are fewer impediments or sources of resistance to the free flow of electrons.

There are limits, nonetheless, to how resistance-free normal conductors can become. Due to various kinds of defects in the lattices of normal conducting materials, resistance to electron flow will be present even when such materials are lowered to near absolute zero – that is, approaching -273 degrees Celsius, or there about.

Something different happens in the case of superconductors. To simplify – hopefully not overly so -- a much more complex process, electrons tend to pair up below a certain critical temperature, and in the process, resistance to electron flow also disappears.

The transition from the presence of resistance to the absence of resistance is considered to constitute an example of a spontaneously broken symmetry. Nambu argued that something of a similar nature might have taken place either prior to or during the Big Bang ... that is, fundamental particles -- which, initially, were massless -- transitioned to a state that exhibited mass.

Quantum Queries

Nambu's theory concerning the origins of mass predicted that, in addition to the emergence of mass, there would be massless, scalar particles or excitations that also would arise under the conditions of symmetry breaking. However, the proposed existence of those particles – sometimes referred to as Nambu-Goldstone particles -- was problematic.

For example, as indicated in the foregoing paragraph, Nambu-Goldstone bosons were described as being massless. Since they were hypothesized to be massless, very little energy would be necessary to generate them, and, yet, no one had ever detected the presence of those kinds of particles during any high-energy experiment, nor under any other set of experimental conditions in which such particles might have been hypothesized to arise.

Most physicists concluded that the idea of a spontaneous symmetry breaking process that would give expression to the generation of particle masses at the beginning of the universe was inherently problematic because of the apparent need to suppose that the massless Nambu-Goldstone particles also had to be present during the foregoing sort of process. Yet, there was no evidence to indicate that such particles existed.

Philip Anderson (the "A" in the preceding amalgamation of letters listed near the top of page 114) noted in 1963 that the presence of massless particles during the process of spontaneous symmetry breaking wasn't necessarily the problem that many physicists considered it to be in relation to Nambu's ideas concerning the origins of mass in fundamental particles. Anderson indicated that during the process of symmetry breaking in superconductors (when resistance to the flow of electrons disappeared), massless particles temporarily made an appearance and, then, those particles acquired mass as a result of the ensuing dynamic that led to the emergence of superconductivity.

Photons, which are massless, are described as becoming massive when a potential superconductor transitions to becoming an active superconductor. This is known as the Meissner effect.

A number of theorists followed up on the foregoing ideas of Nambu and Anderson and tried to suggest possible ways for resolving some of the problems surrounding the issue. They did so through the pages of the *Physics Letters* journal published by CERN, and, as is often the case with respect to such journals, someone sent in a reply that was critical of the ideas being proposed.

Higgs was following the thread of the foregoing argument. He decided to weigh in on the discussion.

He wrote a paper (consisting of just 79 lines of text material) that advanced a modified version of Nambu's ideas which purported to account – at least theoretically -- for how a massive particle, with zero spin, might arise when a certain condition of symmetry (i.e., all fundamental particles are in massless state) was broken spontaneously. The aforementioned framework of understanding came to be known as the "Higgs mechanism".

Higgs paper was rejected. In fact, as Higgs subsequently discovered, not only was his paper rejected, but, as well, the editor of the *Physics Letters* publication indicated to various people that Higgs' ideas were irrelevant to particle physics.

In light of the Nobel Prize that would be awarded to Higgs some 50 years later for the ideas put forth in the foregoing paper (together with an idea that was introduced through a couple more paragraphs that were added after the fact of his initial paper's submission), one might note in passing how fragile the communication process is through which ideas are either permitted to be shared with other scientists or prevented from reaching the light of day in that respect. I have no idea how intelligent or knowledgeable the person was who served as the editor for *Physics Letters* at the time when Higgs submitted his first paper concerning the origins of mass, but in the hindsight permitted by the subsequent events of history, clearly that editor didn't necessarily have much understanding about what was, and was not, relevant to particle physics, and, yet, that editor was, to some extent, making judgments about what ideas scientists (both theoretical and experimental) could, and couldn't consider.

Fortunately, there was more than a single way through which to place one's ideas before the minds of other scientists. Higgs reviewed his paper, decided to add several more paragraphs, and in the last paragraph he introduced the idea of the particle that later would bear his name ... the particle that bore certain quantum properties that would constitute the tell-tale sign indicating that an underlying field existed through which mass might be conferred upon fundamental, massless particles during the process of spontaneous symmetry breaking.

This time, Higgs updated paper was submitted to a journal that was published in the United States rather than through CERN – namely, *Physical Review Letters*. His paper was conditionally accepted for publication ... Higgs needed to add one item to his paper.

More specifically, Higgs needed to make reference in his paper to another paper that was to be published by the journal on the very day that the latter journal had received Higgs' paper. The paper to be referenced by Higgs was written by two Belgian physicists, Robert Brout and François Englert.

The Broout/Englert paper dealt with the same issue Higgs was writing about – the origins of mass -- but the two foregoing authors had approached the problem from a different direction. However, Higgs' paper was the only one of the two papers to introduce the idea of a boson with certain properties – later to be known as the Higgs particle – which played a telltale role in the process of symmetry breaking that led to the emergence of mass.

The Higgs particle is an exchange particle – a boson -- that carries the potential to generate mass when it interacts with particles that are receptive to such a capacity. While several other basic boson particles (e.g., photon, gluon, and the hypothetical graviton) are massless, the Higgs boson was hypothesized to have mass.

However, contrary to Higgs' aforementioned belief about how things began – that is, through spontaneous symmetry breaking -- symmetry doesn't necessarily have to be broken in order for a Higgs-like field, or a Higgs-like boson, or a Higgs-like mechanism to exist … although there might have to be some adjustments made to the foregoing concepts within a context in which symmetry was not broken. What is essential is that Higgs predicted the possible existence of a massive particle with zero spin that could arise from a certain kind of scalar field and, in the process, the field would have the capacity to confer mass on certain kinds of particles that interacted with that field.

According to Higgs, elementary fermion particles (particles with half-integer spin that obey the Pauli exclusion principle and are not part of a composite complex of some kind) started out massless. Then, during a process of spontaneous symmetry breaking -- which either helped set the Big Bang in motion or occurred during that event -- an already existing (but previously dormant field) began to interact with certain fermion particles that initially had been massless and conferred mass on those particles – at least to a degree.

Since the field in question apparently existed prior to, and, therefore, independently of the Big Bang, one also could hypothesize that the Higgs field might never have been dormant and could have been operational even if the Big Bang did not take place (and I will have more to say on the issue of the Big Bang in the next volume when various facets of cosmology are explored). If the foregoing possibility were true, then, those particles that are susceptible to the presence of the Higgs field might always have exhibited mass.

In other words, perhaps the foregoing sorts of particles were never in a massless condition because the Higgs field – possibly -- was never dormant. Whether, or not, the universe began with fundamental particles that exhibited mass, or those particles only subsequently acquired mass (one picosecond later), might affect how the universe unfolded if it did, in fact, begin with a Big Bang.

The so-called Standard Model of quantum mechanics indicates that there are 24 basic forms of matter in nature. Six of those fundamental building blocks consist of leptons (electron, muon, and tau particles, plus their associated neutrinos), while the remaining 18 building blocks consist of quarks.

There are six kinds of quarks. These go by the names of: up, down, top, bottom, charm, and strange.

Each of the foregoing quarks comes in three editions depending on what kind of charge is present. The charge varieties are referred to by the color terms: 'blue', 'green', and 'red', but these designations are arbitrary and are more about being able, through an adopted convention, to differentiate among the three kinds of charges than it is a matter of charge being a function of color in any sensory or physical sense.

Quarks exhibiting properties that indicate the presence of a given kind of color charge are attracted to other quarks with different color charges. Moreover, just as every other kind of particle has a partner from the 'dark-side', so too, each quark has an antiquark associated with it.

There are a number of other particles described through the Standard Model that are not considered to be material building blocks, but, nonetheless, have a role to play in making the material world possible. Four of these other kinds of particles are referred to as "bosons", and they are considered to carry one, or another, kind of force or property that is mediated through them in relation to other kinds of particles that are open to being part of such a mediated dynamic of exchange or conferral process concerning a given kind of force or property.

For example, the photon is said to carry the electromagnetic force and, thereby, mediates the exchange of, or conferring of, electromagnetic force. The gluon is described as carrying the strong force that mediates the exchange of that property among quarks and helps hold protons or neutrons together), while W (both positive and negative) and Z bosons have been identified as carriers or mediators of the weak force (which is involved in, among other things, radioactive decay as well as the phenomenon of one kind of quark changing into another kind of quark through the exchange of a W particle).

The graviton is a hypothetical boson. It is believed to be the carrier of gravitational force and, therefore, the mediator of exchanges involving gravitational effects, but as the foregoing use of the term "hypothetical" suggests, the graviton has not, yet, been discovered.

The Higgs particle is a fifth kind of boson, and, as previously noted the Higgs is involved in mediating, exchanging, or conferring the property of mass. Until 2012, the ontological status of the Higgs boson was like that of the graviton ... i.e., hypothetical in nature.

It is important to understand that not all particles that have mass owe the entirety of their mass to the existence of the Higgs field. Rather, the Higgs field provides a way of accounting for how certain particles (for example, leptons ... that is, electron, muon, tau particles and their respective neutrinos) which otherwise would not possess mass are capable of exhibiting mass in the presence of the Higgs field.

Certain particles – for example, protons and neutrons -- give expression to the presence of mass, but only a small portion of that mass is due to the presence of the Higgs field. The majority of the mass of protons and neutrons is generated through the dynamics of quarks and gluons that bind quark triads together within protons and neutrons.

Nonetheless, exactly how mass arises through the dynamics of gluons is, to some extent, a question that has not been fully answered. Conceivably, just as some theorists have hypothesized the existence of a Higgs-like mechanism that is different from the "usual" Higgs mechanism and is alluded to – despite its unknown nature -- as a possible way to account for the generation of the mass in right-handed neutrinos – if the latter exist – so too, there might be some kind of an additional Higgs-like field to account for why gluons have the mass they do, or, perhaps, the mass of gluons is a yet-to-be-discovered function of the Higgs field that was uncovered in 2012.

The role of gluons in the generation of mass is somewhat complicated. Unlike photons which do not appear to have a charge associated with them, gluons are said to be carriers of the color charge that are exchanged among quarks and help keep the internal structure of protons or neutrons tied together, and, as a result, gluons not only mediate the strong force but, as well, those boson particles participate in the strong force, thereby making the nature of gluon dynamics a lot more difficult to analyze than is the role played by photons in the electromagnetic force (although the latter role is, by no means, simple, and an accurate way to describe those transactions took time to develop).

Many theoretical physicists have acknowledged that current mathematical treatments of the Standard Model allow for the possibility that more than one kind of Higgs field could exist. The experiments that were concluded in 2013 at the LHC run by CERN did not appear to have generated data that, to date at least, demonstrated such additional Higgs-like fields can't or don't exist.

However, additional work will be done with respect to the foregoing issues (and others) during the next series of experiments that are planned to begin at CERN sometime in early 2015. What the world of physics will look like once the data from those experiments have been exhaustively analyzed and interpreted remains to be seen.

There are a variety of questions that need to be answered and issues that need to be settled. The questions I have in mind allude to facets of reality that might, or might not, exceed whatever is discovered during the CERN experiments slated to begin in 2015.

For instance, bosons are said to carry forces and properties that are conferred on, or exchanged with, those particles that are receptive to the presence of such forces and properties. What, exactly, is involved in the dynamics of mediation, exchange, conferral, or carrying of forces (electromagnetic, strong, weak, and gravitational) or properties (mass) is not clear.

Many of the behavioral aspects of the foregoing sorts of dynamics have been determined with considerable precision. Physicists know how to measure and quantify the properties that are generated through those sorts of forces or properties. In addition, physicists are able to predict what will happen quantitatively and behaviorally when those kinds of mediation, exchange, or conferral processes take place in relation to particles that are receptive to such mediation/exchange/conferral properties.

Nonetheless, physicists don't appear to understand precisely how bosons carry forces or properties. Nor do scientists seem to understand the precise character of the exchange/mediation/conferral process.

The surface properties of those processes have been quantified. However, the underlying dynamics of those processes are still something of a mystery ... assuming, of course, that such dynamics exist.

For example, how does a boson carry any given force or property that is to be exchanged or conferred under the appropriate circumstances of mediation? Does the boson carry such a force/property on or within itself and, then, at the appropriate time, releases or generates that force/property, or, alternatively, does the boson merely serve as a locus of manifestation through which an underlying field of some kind gives expression to that force or property?

In both of the foregoing cases, there are further questions that need to be asked. If the boson generates, or carries, or mediates a given force or property, then, how does it do so? On the other hand, if the boson is merely the juncture through which the underlying field gives expression to itself (and that juncture is referred to as a boson particle), then what is the specific character of the dynamic that causes the field to express itself the way it does at that given instance of manifestation?

Mathematical field theories have the capacity to capture the surface, behavioral results of the foregoing sorts of underlying dynamics. However, those same field theories do not appear to be able to account for how those results are generated … just that they are generated and that the behavioral products of boson and/or field dynamics have certain measurable, quantitative characteristics.

When one believes that reality is just a matter of random, inherently uncertain, probabilistic events (and nothing more), then, one does not need to worry about the foregoing issues. One merely latches onto surface, behavioral, measurable, quantitative results whenever and wherever they can be found and, then, proceeds to claim that those outcomes are all that can be known … just do the calculations that are needed to solve the kinds of problems that involve the prediction of certain quantities and/or how those quantities will unfold over time in a given set of circumstances

Such a claim is never substantiated. It only constitutes the presumptive and operational framework through which, or within which, various quantities are described, predicted and calculated.

| Quantum Queries |

How does a force "know" that particles are receptive to it? How does a particle "know" that it is receptive to a given force or property?

Of course, one could argue that no "knowledge" is involved. Particles – whether bosons or fermions – just do what comes naturally to them.

Nonetheless, one would like to know if the dynamic involving the conferring, exchanging, or mediating of a given force or property is one-sided or two-sided. In other words, is a given fermion receptive to the force or property associated with a particular boson because the structural properties of a fermion render the latter receptive, in some way, to the properties of the former particle?

If so, then, bosons don't necessarily just mediate, confer, or exchange a force or property in relation to receptive fermions. Bosons and fermions would be engaged in an interaction of some kind in which both sides bring something to the dynamic.

Certain bosons might be carriers of a force or property. However, the structure and quantum state of the would-be recipient might determine whether, or to what extent, or in what way, the force or property gets conferred or exchanged. As such, fermions would not be passive recipients of forces or properties that are being mediated by a boson of some kind, but, rather, fermions would be actively engaging bosons, much like molecules of the right shape and charge interact with appropriately shaped and charged membrane molecules of a cell.

For instance, one can ask: Does the Higgs field confer mass on a particle? Or, one can ask: Are fermions so structured that they are receptive to interacting with the Higgs field in a way that results in mass becoming associated with such fermion particles?

How does the Higgs field differentiate between the masses that are to be conferred on different particles with different properties that will lead to the manifestation of different masses? The Higgs field can't interact with different fermions in the same way but, rather, there would seem to have to be properties within, or associated with, a fermion that shapes the way and extent to which mass can be conferred.

For instance, according to many physicists (and this line of thinking started, for the most part, with the work of Steven Weinberg, Abdus Salam, and Sheldon Glashow), W and Z particles are hypothesized to have begun interacting with the dormant Higgs field following some form of spontaneous symmetry breaking that took place prior to, or during, the Big Bang (and the precise manner in which this occurs is currently unknown). During the foregoing interaction, W and Z particles are hypothesized to, somehow, acquire a longitudinal dimension to the wave associated with those particles to go with its already existing transverse oscillation (which is perpendicular to the plane along which a particle is moving).

The aforementioned longitudinal component enables those waves to oscillate in the direction of travel. As the W and Z particles plow through the Higgs field with such a longitudinal form of oscillation, mass is generated, however, the emergence of mass might require more than the presence of such a longitudinal wave.

The basic Higgs field consists of two charged components and two neutral components. Collectively, this is all referred to as being a complex field.

The two charged components of that complex Higgs field interact with the positively and negatively charged W particles and, together with the acquired longitudinal component of the wave associated with the W particle, give rise to mass in the two editions of the W boson.

One of the two neutral components of the complex Higgs field interacts with the Z boson, and from that dynamic, the Z particle acquires mass. The other neutral component of the complex

Higgs field constitutes the Higgs boson ... it is what is left over after mass has been conferred, mediated, or exchanged.

How the longitudinal component of the waves associated with W and Z particles arises is not, yet, known. How the acquisition of such a wave property interacts with the aforementioned charged and neutral components of the complex Higgs field and, thereby, leads to the generation of mass as the W and Z 'wavicles' interact with the Higgs field is not, yet, known.

The foregoing hypothesis might, or might, not extend to the realm of neutrinos. Left-handed neutrinos and right-handed neutrinos – if the latter exist – could interact with the Higgs field in different ways, and it is possible that different kinds of Higgs fields might be involved with left-handed and right-handed neutrinos.

A further complication might arise in conjunction with leptons (electron, muon, and tau particles, together with their respective neutrinos) and quarks. Currently, theoreticians believe that prior to engaging a Higgs field, quarks and leptons exist in what is referred to as a single spin state that either does, or does not, spin in the same direction in which those particles are moving.

However, when quarks and leptons engage a Higgs field, or such a field awakens from its dormant state following some kind of spontaneous symmetry breaking prior to or during the Big Bang, then, according to some theorists, quarks and leptons acquire the capacity to exhibit both sorts of spin states, and, in the process, mass arises. Whether, or not, more than one kind of Higgs field is involved in the foregoing dynamics, and exactly how the second spin state is acquired, and how having both spin states will give rise to mass are issues that need to be determined experimentally (perhaps in 2015).

Another set of problems involving mass arises in conjunction with the fact that the 24 fundamental building blocks of nature have been partitioned into three generations of particles that, supposedly, are like the other generations in every respect except for mass. For instance, there are electron-like leptons (muon and tau particles) and electron neutrino-like leptons (muon neutrino and tau neutrino) existing within each generation, but the masses of the particles in each succeeding generation become heavier.

To add more mystery to the matter, only one of those generations of particles – the first generation – tends to be observed in nature. In other words, aside from the specialized and somewhat artificial confines of colliders and accelerators, for the most part, only the first generation of fundamental building blocks tend to be observed in nature.

Second and third generation elementary building block particles seem to restrict their activities to the precincts of high-energy physics apparatuses and experiments. However, there are some exceptions to the foregoing tendencies.

The first generation of particles (the one that tends to dominate the dynamics of those portions of the universe that take place beyond the horizons of man-made accelerators and colliders) includes: electrons and the electron neutrino, together with three editions of 'colored' up and down quarks. The second generation of particles includes: the muon and the muon neutrino, along with three color-charged versions of charm and strange quarks, and, finally, the third generation of particles consists of: the tau particle and the tau neutrino, as well as the three color charges for top and bottom quarks.

As indicated earlier, the primary difference between electrons and their second and third generation cousins (muon and tau particles respectively) is said to be just one. Each succeeding generation is heavier than the one before it.

Similarly the difference between succeeding generations of quarks is supposedly just a matter of mass. Earlier generations of quarks have less mass than succeeding generations do.

| Quantum Queries |

There are, at least, several questions that come to mind in relation to the foregoing three generations of particles. Why, for the most part (and neutrinos appear to be at least one set of exceptions to such a general rule), do only first generation leptons (electrons and electron neutrinos), together with quarks (three color-charged editions of up and down particles), tend to show up in nature – that is, outside of high-energy labs?

Secondly, why do material building blocks (e.g., electron, muon, and tau particles) that are supposedly the same in every respect exhibit different masses when they engage one, or another, Higgs field? What sort of longitudinal wave component, spin property, and/or other kind of quantum feature is responsible for helping to give rise to a different kind of mass when that property, component, or feature engages one, or another, Higgs field, and, if such a property exists, then, can one actually say that the only differentiating feature among the three generations of material building blocks necessarily is just a matter of mass?

Furthermore – and returning, momentarily, to the opening sections of this chapter -- if one considers the rest mass of an electron, then one can't help but notice that such a quantity involves a very precise figure: $9.10938291 \times 10^{-31}$ kilograms. What is the nature of the interaction between an electron and a Higgs field that would generate that kind of a precise value in such a constant fashion … a constant that carries so many ramifications for, among other things, how – or if – various kinds of chemical dynamics will take place in the universe?

The relationship between, on the one hand, a theory or hypothesis and, on the other hand, the use and acceptance of that theory or hypothesis can be very complex. There are many currents with which to contend … both technical and social.

For example, as indicated previously, the editor of the *Physics Letters* journal associated with CERN had dismissed Higgs ideas as being irrelevant to particle physics. The ideas of two other theorists, Gerald Guralnik and C. R. Hagen (who, along with Tom Kibble, had developed their own approach to the idea of spontaneous symmetry breaking and the origins of mass), were also dismissed by a person with considerable stature in the world of physics ... the Nobel laureate, Werner Heisenberg.

In the summer of 1965, Heisenberg organized a small gathering of physicists. Guralnik and Hagen were interested in talking about their ideas concerning spontaneous symmetry breaking in relation to the origins of mass issue, and, they were given the opportunity to do so at Heisenberg's event.

The general reception of the conference's participants with respect to the talks by Guralnik and Hagen was less than underwhelming. Heisenberg referred to their ideas as "junk" ... which only goes to demonstrate that being right about some things in the past doesn't automatically mean that one will be right about everything in the future.

Critical reflection is an important component of the scientific process. However, there are a great many differences between, on the one hand, a thoughtless, abrupt dismissal of an idea, and, on the other hand, a rejection of ideas that occurs in a collaborative context of constructive criticism that seeks to assist everyone to work toward the truth.

Some scientists look at the process of denigrating and picking apart ideas – whether written or spoken -- as integral to making progress in science. Unfortunately, such a process – when it contains elements of blind acceptance of one's own ideas and a desire to control how other people think -- might be more a function of ego and ideology than it gives expression to an open and sincere search for the truth.

When Heisenberg referred to the ideas of Guralnik and Hagen as "junk", Heisenberg, apparently, had forgotten how he felt when Bohr reduced him to tears as Bohr was rejecting Heisenberg's approach to the uncertainty principle. Or, alternatively, perhaps,

Heisenberg had learned too well from the intellectual bullying that Bohr had exhibited toward Heisenberg (and, later, toward Schrödinger when the latter individual was sick, and, yet, Bohr wouldn't leave Schrödinger alone until Schrödinger capitulated to Bohr's perspective).

Steven Weinberg experienced a different side of the dynamic between theory and use/acceptance than Guralnik and Hagen did. In 1967, three years after Higgs had written his paper, Steven Weinberg had been trying to use the Higgs mechanism to account for certain differences between neutrons and protons.

The Higgs particle had not, yet, been found. There was no really good reason for Weinberg to use the Higgs mechanism other than that it had something to do with spontaneous symmetry breaking, and, as well, he was interested to see where the Higgs mechanism might lead him with respect to being able to differentiate, in certain subtle ways, between neutrons and protons.

He was getting nowhere fast until he realized that the nonsensical results he was coming up with had nothing to do with the differences between neutrons and protons. Instead, what he had been fooling around with seemed to point in the direction of the inter-relationships between weak and electromagnetic forces ... inter-relationships that have since come to be known as the electroweak force.

Weinberg believed that at some point prior to, and/or shortly after, the Big Bang, both the electromagnetic force and the weak force were unified in some sense. However, as the universe cooled, the two forces became separated, and the cause of the division was the Higgs mechanism.

If Weinberg's new theory was correct, it indicated there were three new particles that, at some point, should turn up in accelerators and/or colliders. Those particles were the positive and negative W (i.e., weak) bosons, together with the Z (zero electrical charge) boson.

The Higgs mechanism did not affect photons, and therefore, those particles remained massless. Consequently, the electromagnetic force that was separated off from the unified force that had existed prior to, or shortly after, the Big Bang, was able to project its strength (via the photon) across great distances at the speed of light, while the weak force that separated off from the previously unified force – also through the Higgs mechanism -- was mediated by particles (the two W's and the Z) that became massive through the Higgs mechanism and, as a result, the effective sphere of influence of those bosons was very limited as far as distance is concerned.

Six years earlier -- in 1961 -- Sheldon Glashow had developed a theory that integrated weak and electromagnetic forces. However, because the Higgs mechanism was still three years in the future, Glashow didn't have the tool that Weinberg had – i.e., the Higgs mechanism – to make Glashow's model work properly.

At least two broad questions arose in conjunction with Weinberg's work. Firstly, did the particles (two W's and a Z) that were predicted by his theory actually exist, and, secondly, could one find a way around the infinities that sometimes arose when calculations were made in conjunction with that theory.

The first question was answered through the work of, among others, Carlo Rubbia and Simon van der Meer in 1983. As far as the second issue noted above was concerned, Richard Feynman had devised a mathematical technique referred to as "renormalization" that helped to rid calculations of such infinities in the context of quantum electrodynamics.

Perhaps, something of a similar nature could be done in relation to electro-weak dynamics. Of course, there still was no direct evidence that the Higgs mechanism, Higgs field, or Higgs particle even existed, but, nonetheless, if nothing else, the foregoing triad of ideas had great heuristic value because they had helped Weinberg to make a very important breakthrough concerning two important themes of nature ... the electromagnetic and weak forces.

In 1970, Martinus Veltman and Geradus 't Hooft, working at Utrecht University in the Netherlands, developed a renormalization-like technique for ridding calculations of infinities in theories like the one devised by Weinberg that involved, at least in part, particles with considerable mass. Although infinities did arise when calculating the nature of interactions, those infinities immediately cancelled out one another.

't Hooft had been trying to develop, from first principles, a theory involving Weinberg-like theories. As he did so, he came to understand that the Higgs mechanism was at the heart of that theory ... in fact, 't Hooft had been able to derive Higgs theory while approaching the issue from an entirely different theoretical direction than Higgs had done.

Veltman and 't Hooft had been engaging the problem of infinities from somewhat different perspectives from one another. When Veltman included the Higgs mechanism in his own model, as 't Hooft had been able to do, Veltman was able to generate sensible calculations.

The work of Veltman and 't Hooft lent mathematical validation to the work of Weinberg and Higgs, and, in the process, assuaged the anxieties that many scientists harbored concerning the issue of unwanted infinities. Nonetheless, while devising the foregoing sort of techniques might make mathematical sense, this still leaves unanswered questions about what such techniques have to do with the ontological dynamics to which those mathematical techniques supposedly allude.

Before discussing the infinities problem with 't Hooft, Veltman felt that the Higgs mechanism was nothing more than a theoretical trick that enabled one to arrive at certain kinds of solutions. After exploring the issue with 't Hooft and, then, incorporating the Higgs mechanism into his own model, Veltman was able to generate sensible answers, and, as a result, Veltman came to understand the scientific value of that mechanism.

However, wasn't the renormalization-like technique that Veltman and 't Hooft developed something of a trick itself? Irrespective of whether, or not, one considered infinities to be inherent in the nature of ontology, there were problems.

On the one hand, if ontology did not contain infinities, then, both the equations and calculations that gave rise to those infinities as well as the techniques that were developed for removing them (by Feynman, Veltman, and 't Hooft), were, obviously, not describing reality but, instead, those calculations and equations constituted a way of engaging reality through filters that both introduced, and, then, removed infinities. On the other hand, if reality did contain infinities, then, ontologically speaking, what did it mean for such infinities to cancel out one another and what were the actual dynamics for such a process of cancellation?

The foregoing issues notwithstanding, there is another problem associated with the Higgs mechanism. While Weinberg, Veltman, and 't Hooft – each in his own way – were demonstrating that the Higgs mechanism seemed to play a central role in quantum physics, and, quite possibly, in the way the universe might have come into being, nevertheless, no one knew how and where to look for the Higgs boson because no one knew its mass.

For decades, scientists had sought evidence indicating that such a particle was real. Its discovery was crucial because if it existed, then it gave expression to a physical marker that constituted evidence indicating that the underlying Higgs field through which the Higgs mechanism conferred mass on particles actually existed.

However, there were many technical problems to overcome with respect to the hunt for the Higgs particle. First of all, the Higgs boson -- if it existed -- was hypothesized to be very unstable, flicking into and out of existence in one hundred trillionths of a trillionth of a second and, therefore, the collider detectors would have to be calibrated to very fine tolerances to even be able to catch sight of the debris field left by a decaying Higgs particle.

Secondly, the Higgs particle had been hypothesized as being likely to show up in the presence of a Z particle. Since scientists believed that the – at that time -- hypothetical Higgs particles generated a shower of decay debris – just as the Z particle did -- one would have to sort out the two sets of debris fields by trying to determine from which particle – the Higgs or the Z boson, along with other possibilities – such decay debris might have arisen.

Approximately 70% of the time, Z particles decayed into various quarks. However, nearly 20% of the time, Z particles decayed into a stream of neutrinos (whose presence can be inferred on the basis of the small amount of energy that is carried away in the form of those neutrinos). The other 10% of the time, Z particles generated a debris field consisting of electrons or muons (a heavier edition of electrons).

Higgs bosons were predicted to generate a similar kind of debris field. Plus, there were other decaying particles that could generate debris that would be difficult to distinguish from what might happen with a Higgs particle.

Consequently, it was hard to know what one was observing in any given instance. The chances of encountering a false sigma – in other words, data that suggested a Higgs boson might be present when this was not the case – had to be taken into consideration.

Finally, the discovery of new particles like the Higgs is not a matter of capturing evidence concerning just one event. Rather, the existence of such particles is based on the statistical analysis of a number of events.

More specifically, one compares what is observed against what one is likely to see if such a particle didn't exist. If one arrives at a calculation that indicates there is less than one chance in a million (referred to as a 5 sigma – standard deviation – level of confidence) that the observed events are not due to the presence of the sought for particle, then, one has what is considered by most scientists to constitute fairly reliable evidence that a certain kind of particle discovery has been made.

Toward the end of the series of experimental runs at CERN's LEP (Large Electron Positron) collider in 2000, a number of observations were made in conjunction with two different detection systems suggesting that a Higgs boson might have been present during those events. One detector produced data that, when analyzed, yielded a 3.9 sigma level of confidence concerning the likelihood that a Higgs particle had been discovered, while the other detector's data, when analyzed, only rose to a level of confidence of 2.7 sigma.

The foregoing numbers looked promising as far as indicating the possible detection of a Higgs particle is concerned. Nonetheless, those numbers fell short of the gold standard of confidence ... namely, 5 sigma.

Considerations of additional financial and political costs, as well as significant delays in constructing the proposed LHC that was to pick up where the LEP collider experiments had left off, were weighed against whether, or not, the foregoing sigma levels (3.9 and 2.7) were worth the gamble of continuing on with further experiments in 2001 in the hopes of generating the data needed to pin down the existence of the Higgs boson at a higher level of confidence.

The risk-reward calculus was decided in a way that was more concerned with risks than with rewards. The plug was pulled on

LEP in November of 2000, and physicists would have to await the construction of the LHC before they could continue to chase the elusive Higgs.

There is a potentially interesting issue that lingers on in conjunction with the foregoing sigma results that arose toward the end of the LEP collider runs in 2000. The issue is one that might, or might not, be cleared up when CERN begins its new series of experiments early in 2015, but it didn't seem to be resolved during the series of experiments at CERN that ended in 2013.

More specifically, the mass of a particle will, in part, indicate how it might have formed, and, as well, how it might decay. For some time, scientists have hypothesized that the most probable manner through which Higgs particles will arise in a collider is when two gluons (responsible for keeping quarks tethered to one another within protons and neutrons) are fused together after having been slammed into one another.

Prior to 2012 – that is, before scientists knew what the mass of the Higgs boson was (or, at least, one of them) -- physicists had to search for different signature profiles in the decay debris that would vary as a function of the mass of the Higgs boson. For example, if the mass of the Higgs particle were more than 130 GeV, the decay products of that particle likely would involve lepton jets of some kind (electrons, muon and tau particles, plus their respective neutrinos), whereas if the mass of the Higgs particle was around 114 GeV, the debris is likely to consist of gamma rays.

However, the Higgs could also decay into several Z bosons that, in turn, could decay into an electron-antielectron pair or several muons. Alternatively, the Higgs particle might decay into a bottom quark and its antiparticle, and each of the latter possibilities could decay into a jet of hadrons made from quarks of one description or another.

Finally, a decaying Higgs boson might generate two W bosons. These W bosons could, in turn, decay into a muon or an electron-antielectron pair, plus a neutrino-antineutrino pair.

Unfortunately, all of the foregoing possibilities might also arise through other decay mechanisms that are taking place amidst the hundreds of millions of proton-proton collisions that are occurring in the LHC with each passing second. Consequently, figuring out what one is dealing with can be a very difficult problem.

On June 15, 2000 some sort of an event took place in conjunction with the Aleph detector for the LEP collider. The event consisted of four jets of particles.

Two of the jets seemed to involve quarks that had been produced during the decay of a particle with a mass of approximately 91 GeV. That decaying particle appeared to bear the signature of a Z boson.

The other two quark jets came from the decay of a particle that was much heavier than a Z particle. Those quarks were associated with a particle whose mass was about 114 GeV.

Several weeks after the June 15, 2000 particle made its presence known through the Aleph detector, another event occurred that exhibited approximately the same mass (114 GeV) as the aforementioned Aleph detector particle had but through a different detector, L3. The fact that a different detection system had registered the presence of an unknown particle that was similar to the one detected through Aleph, seemed to indicate that whatever had taken place in the two detectors was not due to some sort of detector glitch since it was extremely unlikely that two different detection systems would break down in the same way within weeks of one another.

Although the masses of the foregoing two particles were approximately the same, nonetheless, in other respects, they displayed different properties from one another. However, such differences were consistent with what one might find with Higgs-like particles.

If the LEP experiments had continued on into 2001 (as some scientists wanted to do), the identity of the foregoing particles might have been resolved. Nonetheless, those experiments were not conducted, and, therefore, the identities of the two, aforementioned particles (together with a few other Higgs-like candidates that surfaced in the Aleph and Delphi detectors some time after June 15, 2000) were never established.

On July 4, 2012, CERN announced that sufficient evidence had been compiled in conjunction with the experiments being run through the LHC (Large Hadron Collider) to indicate that at least one form of Higgs boson did, in fact, exist. As noted previously, the Higgs particle had a mass of approximately 126 GeV.

The foregoing Higgs mass is 12 GeV greater than what had been detected in 2000 through the LEP. If the latter LEP particles were not Higgs bosons, what were they?

Were the 2000 events giving expression to known particles that, for some unknown reason, were displaying anomalous properties? Or, were those events connected, in some way, to new particles that had not, yet, been identified … perhaps, connected to Supersymmetry?

Obviously, the story does not end with the announcement that a Higgs boson had been detected. Throughout the pages of this chapter, a number of problems and unanswered questions have been introduced that arise in conjunction with the issue of the Higgs field, the Higgs mechanism, and the Higgs boson.

The problems and questions being alluded to in the foregoing paragraph might, or might not, be answered satisfactorily through the series of experiments that will be run at CERN over the next several years. Peter Higgs feels there are a number of different kinds of Higgs-like particles yet to be discovered, and, of course, each particle constitutes a physical marker for some kind of underlying field.

| Quantum Queries |

Moreover, Higgs believes that the mathematics of one or another version of Supersymmetry will be able to integrate all of the foregoing particles and fields into a consistent and viable account that will extend the explanatory horizons of the Standard Model of quantum physics beyond its present capabilities. Higgs' vision might, once again, prove to be correct, but, thus far, and as previously noted, none of the particles predicted by Supersymmetry (and there are many of these hypothetical entities in any given model of Supersymmetry) have, yet, to be discovered.

The Higgs boson that has been detected at CERN weighs in at around 126 GeV. This fact, in itself, seems to have eliminated quite a few versions of Supersymmetry from consideration because – and quite apart from the fact not one of the predicted Supersymmetry particles have turned up thus far in experiments – those models were based, in part, on the premise that the Higgs boson was much heavier than it apparently is.

Perhaps, future experiments will discover the existence of other Higgs bosons possessing a mass that is more compatible with the foregoing Supersymmetry models (There will be more discussion involving Supersymmetry in Chapter 8). Or, maybe, future discoveries will eliminate all but one Supersymmetry model ... or, perhaps, forthcoming data will indicate that all Supersymmetry models are problematic in one, or more, ways.

Before leaving the topic of the Higgs field/mechanism/particle, there is one further issue to be considered. This involves a discrepancy between theoretical calculations and what has been determined experimentally.

This discrepancy arises in conjunction with the dynamic that is said to take place between, on the one hand, the virtual particles that, allegedly, are blinking into and out of existence within the vacuum, and, on the other hand, the Higgs boson. If one calculates how those virtual particles interact with a Higgs boson and, in the process, one adds that total to the mass of the Higgs boson, the figure one comes up with is 1×10^{15} greater than what is indicated by various experimental results.

Unfortunately, if a Higgs particle were as heavy as the foregoing possibility suggests, then, the Higgs boson would not be able to spontaneously break the symmetry of electroweak forces. If such heavy Higgs particles could not break the foregoing sort of symmetry, then, whatever, it is that was discovered at CERN in 2012 could not be the source of mass for elementary particles such as the electron ... some other, much lighter particle would have to help introduce mass into particles like the electron and quark.

One possibility that has bee put forth as a work-around for the foregoing problem is to hypothesize that the dynamic of the virtual particles surrounding a Higgs particle is of such a nature that virtual particles are able to cancel each other out. If this were the case, then, the Higgs particle would be able to remain sufficiently light enough to break the symmetry of electroweak forces and, in the process, separate off the electromagnetic force and the weak force from one another.

However, the foregoing possibility comes with a price tag that seems to strain the credulity of many physicists. More specifically, in order for the above mentioned cancelling out scenario to work, the quantum properties of various forces and particles within the Standard model would have to be fine-tuned to, at a minimum, 15 decimal places, and this, in turn, would require the laws of nature to be extremely sensitive to changes in the values for any of the forces and particles that have a role to play in the Standard model of quantum physics ... possibilities that tend to generate conceptual resistance, if not caution, in many physicists.

Supersymmetry, on the other hand, permits the foregoing canceling out process (involving Higgs bosons and virtual particles) to proceed but accomplishes this in a way that is devoid of any credulity-straining riders being attached to such a solution. Instead, whatever is added to the mass of the Higgs boson through the presence of virtual particles is immediately removed through the presence of the superpartner of that virtual particle.

Although mathematically, things might work out nicely in conjunction with the foregoing process of canceling things out, one still would like to know what the actual dynamical features of the process of cancelation look like. For example, are such superpartners, themselves, virtual particles, and where do the superpartner particles exist (or do they?) prior to the cancellation process, and how do the superpartners "know" that a given kind of non-superpartner virtual particle has become attached to a Higgs boson, and how do such superpartners find their way to the virtual particle in order to cancel things out, and what is the actual nature of the canceling out process?

A few pages back, a mention was made about how the decay products of the Z particle are quarks 70% of the time, and neutrinos 20 percent of the time, and electrons or muons the remaining 10 % of the time. No explanation was given for such a profile because physicists don't know why those events occur in the way they do.

Many quantum physicists believe that those results are just a function of inherently random, uncertain, probabilistic events for which there is no underlying explanation consisting of one, or another, hidden variable. Now, apparently -- at least according to the ideas of Supersymmetry – one is being asked to accept the idea that cancellations involving virtual particles and their 'hypothesized-but-not-yet-discovered' superpartners take place on a regular basis and in a 'just-so' fashion, and, as a result, that dynamic helps make predictive sense of what is actually observed experimentally, and, yet, none of what occurs is supposedly anything more than a set of inherently random, uncertain, and probabilistic events.

Maybe, in the future, some version of Supersymmetry will be proven to be true (or as true as scientific theories ever get).

However, there is a sense about the idea of Supersymmetry that involves a rather convenient fortuitousness that appears to linger in the conceptual air and that resonates – at least to some extent – with the notion of epicycles in the Ptolemaic system of astronomy. In both cases complexities continue to be introduced that, on the one hand, might be able to quantitatively reflect, to varying degrees of accuracy, certain observed results without, on the other hand, necessarily being able to explain what is actually taking place.

Perhaps the vacuum does not contain as much energy as some theorists suppose. Maybe, while there might be some limited number of virtual particles (and their superpartners ... if any) blinking on and off within the vacuum, nonetheless, as pointed out in Chapter 2, perhaps, such activity might be far less than it has been estimated to be by many physicists, and, maybe, there are fewer virtual particles interacting with "regular" particles than some scientists have proposed, and, if so, then, perhaps, there is less of a need for virtual particles and their superpartners to cancel one another out than some physicists have supposed to be the case.

None of the foregoing possibilities does anything to explain the underlying dynamics of such processes ... assuming, of course, that such processes actually take place and that if they do occur, then, they give expression to phenomena that are more than just inherently random, uncertain, and probabilistic events. None of the foregoing possibilities does anything to demonstrate that some version of Supersymmetry does not, in fact, govern certain facets of the laws of the universe.

What is being suggested in the foregoing comments, however, is the following. Perhaps, there is no need to invoke the name of Supersymmetry in order to try to resolve the vacuum energy problem because, just maybe, the relationship between, say, the Higgs particle and the vacuum is a lot simpler than has been supposed, and, as a result, the vacuum energy problem is more of a reflection of problematic theories than it is a reflection of the nature of reality.

If experimental results show one thing, but a theory gives rise to problems (such as the vacuum energy issue) that are not compatible with what has been observed experimentally, then, perhaps, one needs to reconsider one's theory and ask why it is generating values that to not reflect experimental results. In order to resolve the vacuum energy problem, one could hypothesize that there must be some sort of cancelation process taking place between virtual particles and their superpartners in and around Higgs particles, but, one might also hypothesize that the dynamic between virtual particles and superpartners – to whatever extent it takes place at all – is far less extensive than some physicists have supposed and, therefore, there is nothing that necessarily needs to be explained away (i.e., the vacuum energy problem) through invoking some version of Supersymmetry.

Chapter 5: Stringing Us Along

The most fundamental, physical units in string theory are one-dimensional modes of vibrating energy that consist of either small open-ended lengths or closed loops (on a scale of order of 10^{-35} meters). The properties of such segments and/or loops of energy are determined by the nature of the string's oscillation.

Do such entities exist? No one knows!

Is there any reason why the foregoing segments and loops couldn't be smaller – even much smaller – than 10^{-35}? Not really ... the order of scale is somewhat arbitrary.

The means through which energy assumes a one-dimensional form is unknown. The identity of that which constrains strings to be one-dimensional is unknown.

How a given form of oscillation generates all the quantum properties (e.g., charge, spin, mass, etc.) of a string is unknown. What enables oscillations to maintain precise values for certain constants of nature (e.g., Planck's constant, the gravitational constant, the charge of an electron, the speed of light. etc.) is unknown.

Although one might be able to devise various mathematical frameworks through which to describe the vibrational character of a string, the manner in which a one-dimensional string actually vibrates – if such things exist -- is unknown. The identity of that which establishes the parameters within which a string oscillates in one way rather than another way is also not known.

For example, tension refers to a strings resistance to stretching and, as a result, the property of tension is one of the factors that will affect how receptive a string is to different kinds of oscillatory movement. Nonetheless, although Planck scale energy helps determine the character of any given instance of tension, the identity of that which constrains energy and, in the process, generates tension is unknown.

Originally (1968), string theory was intended to serve as a way of accounting for the dynamics of hadrons (particles – such as protons and neutrons -- that are sensitive to the presence of the strong force). However, string theory was unsuccessful as a theory of hadron dynamics, and, eventually, was superseded by a theory involving quarks and gluons.

One of the problems entailed by the initial version of string theory – and, therefore, one of the inducements for scientists to, eventually, look elsewhere for answers concerning the nature of hadron dynamics – revolved the issue of tachyons. Contrary to what some popularized interpretations of quantum physics have supposed, a tachyon does not refer to a faster than light particle, but, instead, indicates the presence of a mathematical instability of some kind within one's theoretical framework.

All the string theories developed at that time were haunted by the tachyon problem, and, therefore, no matter what string theorists did, they couldn't get rid of that defect. The presence of that kind of instability spilled over into, and affected calculations involving, a variety of particles and, in the process, tended to taint the entire theory with a degree of unreliability.

Another problem that couldn't be eliminated from the early versions of string theory was the presence of a spin-2 particle. Yet, despite the ubiquitous presence of such a particle in the mathematics of string theory, there was no empirical evidence to indicate that a spin-2 particle actually existed within the context of hadronic dynamics.

A number of theoretical physicists entered the picture in the 1970s (e.g., Pierre Raymond, John Schwarz, Joel Scherk, and a few others) that developed a Supersymmetrical version of strings known as superstrings. Superstrings had some enticing properties that the previous version of strings did not have.

To begin with, the specter of tachyons had been exorcized from the mathematics of superstrings. Secondly, unlike the first version of string theory, superstrings contained particles exhibiting spin – ½, and this meant that superstrings might be used to model the behavior of ½ spin particles like quarks as well as leptons (e.g., electrons).

Finally, superstring theory was able to give meaning to the spin-2 particle that had befuddled the initial versions of string theory. From the perspective of superstring theory, the spin-2 particle was hypothesized to give expression to the graviton … the entity that was purported to be able to bring together quantum mechanics and general relativity into a unified theory of some kind by accounting for how gravitational force was communicated among objects, both large and small, by means of the exchange of graviton particles.

In the 1980s, John Schwarz and Michael Green developed the idea of superstrings further. Among other things, the two aforementioned theoretical physicists discovered that superstring theory only generated potentially, sensible answers if the mathematical framework consisted of ten dimensions … nine of those dimension were spatial and the remaining one was temporal in character.

The idea of extra dimensions was not a novel concept in string theory. The earlier, hadronic version of string theory had proposed that 26 dimensions would be necessary for describing certain kinds of particle dynamics.

The theory of superstrings had whittled down the number of dimensions required to describe various kinds of dynamics to ten. Nonetheless, because the aforementioned extra dimensions – whether 25 in number or nine in number -- were not visible, various ideas began to be explored to account for the lack of visibility with respect to those additional spatial dimensions.

One of the explanations being alluded to in the foregoing paragraph involved the notion of compactification. This latter term referred to the end result of a process through which additional dimensions curl up in one way, or another, so that while present, they were not readily visible.

Whether, or not, spatial dimensionality is something that – beyond the horizons of mathematics – is capable of becoming compactified is not known. Moreover, even assuming that physical space is capable of becoming compactified, how that process takes place is unknown.

Furthermore, whatever – if any -- extra dimensions might be involved in various kinds of dynamics, there is nothing necessitating that those extra dimensions must be spatial in character. However, if such extra dimensions are, indeed, non-spatial in character, then whatever phenomena are being descriptively represented through the mathematics of extra spatial dimensions must be capable of being preserved accurately when translated into the metrics of some set of non-spatial dimensions.

Another issue concerned with the idea of preservation in string theory involves symmetry. Normally, any symmetry (the preservation of a given property across transformations of one kind or another) that exists in classical theories of physics can be shown to exist, as well, in a quantized version of those classical theories. However, this is not always the case.

For example, when virtual particles are involved, symmetries are sometimes violated. Symmetry violations in quantized theories are referred to as anomalies.

Physicists require all forces to be free of anomalies. In other words, when it comes time to tally up whatever symmetry breaking events might have occurred in conjunction with the presence of one, or another, force, the sum of all those events should add up to zero, and if this is not the case, then, such a framework will generate problematic results with respect to understanding, and making calculations in conjunction with, the dynamics of various kinds of forces.

In 1983, Edward Witten and Luis Alvarez-Gaume had indicated that a similar issue existed in relation to string theory. That is, they maintained that both string theories and quantum field theories were vulnerable to the problem of anomalies.

Schwarz and Green had been able to demonstrate that string theories (of the right vintage) were able to generate mathematical frameworks that -- after all is said and done -- were capable of dealing with anomalies involving symmetry breaking and virtual particles, and, thereby, render such string theories anomaly free and, consequently, capable of leading to sensible calculations.

String theories of the Schwarz and Green variety that consisted of, among other things, just ten dimensions had been able to exorcise tachyons. Such theories also had dispatched the issue of anomalies.

The potential heuristic value – and, therefore, appeal – of new, improved versions of string theory was enhanced further in 1985 by a group of theorists from Princeton (David Gross, Jeff Harvey, Emil Martinec, and Ryan Rohm). That group introduced the notion of heterotic strings.

Closed strings were hypothesized to be able to vibrate both clockwise and counterclockwise. Heterotic strings treated vibrations to the left and right differently from one another, and, as a result, such strings had the descriptive potential to incorporate more kinds of forces within such vibrational degrees of freedom than had been the case in relation to earlier versions of string theory. Moreover, as it turned out, the heterotic strings being introduced by the Princeton group matched the sorts of forces that Schwartz and Green earlier had demonstrated were anomaly free with respect to the presence of virtual particles and the preservation of various kinds of symmetry.

One theoretical hiccup involving the new string theory was connected to the issue of compactification … the curling up of extra spatial dimensions. No matter how theorists tried, they couldn't discover a way to organize their compactified string theories in manner that was capable of accurately describing how the weak force differentially engages left- and right-handed particles.

Indeed, only left-handed particles are sensitive to the presence of the weak force. Yet, compactified theories involving ten dimensions seemed to generate pairs of left- and right-handed particles and, therefore, such compactified theories had no way to account for the left-handed asymmetry that is actually observed in nature with respect to the dynamics of the weak force.

The foregoing problem was resolved in 1985 through the theoretical work of Edward Witten, together with Andy Strominger, Philip Candelas, and Gary Horowitz. Essentially, the foregoing four individuals found a mathematical means that could generate compact manifolds – known as Calabi-Yau manifolds – that were not only capable of differentiating between left-handed and right-handed particles, but, as well, were able to: (a) Filter things through a four-dimensional model (like the world with which we are familiar) that contained 6 additional, compactified dimensions, and (b) preserve the properties of Supersymmetry (This latter model had become increasingly appealing to many theoretical physicists as a way of deepening and extending the Standard Model of physics, and some of its features will be explored a bit more in Chapter 8).

Step by step, the new string theory was building a repertoire of tools, techniques, and concepts that was enhancing its capacity to describe various particles and forces that were found in the Standard Model. At the same time, the same new string theories were avoiding problems such as tachyons and symmetry anomalies that had undermined the initial round of string theories.

One also can add to the foregoing heuristically valuable features the possible potential of updated string theories to unite quantum theory and general relativity through the presence of a spin-2 particle or graviton. When all of these features are put together, the new string theories became very intriguing, and, in the process, such frameworks seemed to allude to a possible Theory of Everything.

The problem with the newer string theories is that they are very long on developing theories that are internally (mathematically) consistent, but they are extremely short on providing evidence – as in: "Virtually none" – concerning the truth of such theories. Furthermore, the likelihood of being able to discover direct evidence – at least in the near term -- that would be capable of demonstrating the correctness of string theories is very small ... although uncovering indirect evidence might be a more feasible undertaking.

The heaviest of the particles that have been experimentally resolved runs around 200 GeV. The mass of some of the predicted string entities are calculated to be about 10^{19} GeV ... which is many orders of magnitude beyond current collider/accelerator capabilities.

Aside from problems involving an almost total absence of empirical evidence to support string theory, as well as, the absence of an experimental wherewithal (both currently as well in the foreseeable future) capable of putting string theory to the test (at least, directly), there was another problematic issue. More specifically, many theorists began to believe that given the right kind of string theory they could derive all of physics from first principles.

For such individuals, physics was no longer a function of empirical activities but, instead, it was a function of pure reason. One could think one's way (at least mathematically) to the truth of things concerning the nature of reality.

The idea that one can grasp the truth of Being through the exercise of mathematics -- which some individuals consider to be the purest and most logical form of reasoning -- is premised on the belief that at its most fundamental level, reality is mathematical in nature. What this means might not necessarily be straightforward, and, as a result, it is a topic that will be explored a little more deeply in volume IV of *Final Jeopardy*.).

The foregoing perspective was captured, to some extent, nearly four centuries ago when Galileo is reported to have said: "Nature's great book is written in mathematical language."

Even if mathematical language can be used to reflect some of the properties of reality, one cannot automatically suppose that reality is a function of language – mathematical or otherwise. A mirror reflects certain facets of reality, but acknowledging such a truth does not necessarily require one to reduce reality down to the properties of a mirror, nor does such a concession require one to suppose that there is nothing more to reality than what can be reflected in such a mirror.

Moreover, languages of many kinds that are not mathematical in nature can be used to heuristically describe various aspects of the dynamics of reality. Poetical, legal, historical, fictional, therapeutic, social, commercial, moral, artistic, political, mythological, and spiritual language all seem to have a capacity to describe various facets of nature in profound – possibly, within limits, even true – ways that are capable of leading to valuable understandings about the potentials inherent in the universe.

Of course, it might be the case that mathematics has a potential to penetrate more deeply into the heart of Being's nature than any other form of understanding and, in the process, reveal that the essence of nature is, indeed, mathematical ... whatever this might mean. On the other hand, mathematics might only be able to give expression to its own mode of descriptive seeing that has a capacity to penetrate down to, or through to, a certain level of order but is unable to see what lies beneath, or beyond, the horizons of its own logical modality of understanding.

In any event, individuals who were proponents of string theory felt that they were achingly close to making, at least, tangential contact with the truth. Yet, what might actually have been taking place (and is still taking place) is that those individuals were on – or are on -- an asymptote path that will never make contact with the essential nature of reality no matter how much progress might be made as far as mathematical considerations are concerned.

All manner of mathematical breakthroughs were being made in relation to string theory. Yet, no one was really sure how, or if, the breakthroughs actually helped to deepen, broaden, or enrich anyone's understanding concerning the nature of reality.

Presently, there are a plethora internally consistent mathematical models (at least 10^{520}) involving strings that are capable of giving expression to a bewildering array of possible: Forms of spontaneous symmetry breaking, particles, forces, dimensions, modes of compactification, vacuum energies, and dynamics. Which, if any, of those models accurately describes the world in which we live is unknown and, in addition, presently there is no reliable set of principles that would enable mathematicians and physicists to be able to identify which of those 10^{520} possible models might best reflect the properties of the universe in which we are living.

One of the few predictions that string theory has made concerns the size or density of the vacuum energy in the universe. The estimates of string theory are much larger than what is actually observed.

A decade, or so, ago, astronomers discovered that the vacuum energy of the universe has a very small but nonzero value. If the vacuum energy were as large as predicted by string theory, then, the universe would have either stretched into nothingness (if positive) or collapsed (if negative) a long time ago.

Yet, neither of the foregoing possibilities has occurred. So, string theory has saddled itself with a sizable problem.

Of course, particle physics fares no better than string theory does when it comes to accounting for why the observed vacuum energy is so small. However, unlike particle physics, string theory is supposed to be describing what occurs on the most fundamental of levels involving the dynamics of gravitons and space, and, yet, quite surprisingly, string theory does not seem to have the slightest clue why the vacuum energy is as small as it is even though such an issue would seem to be situated right in the focal wheelhouse of string theory.

String theory is based on the simplest of possibilities. It revolves about just one feature – namely, the tension on a string that is induced to vibrate.

Nonetheless, string theory has difficulty predicting the character of most facets of the physical universe. While string theory is capable of descriptively capturing some of the broad features of physical reality (e.g., the – at least on the surface – four dimensional nature of physical experience) and, as well, it is capable of incorporating some of the forces and particles of the Standard Model within the context of a string framework, nevertheless, despite more than thirty years of diligent efforts around the world, string theory is unable to reflect many facets of the Standard Model of physics.

Despite the many disappointments that have continued to haunt the early promise of string theory, exploratory activity has not ceased. The framework through which strings are now engaged has expanded from one dimension to many dimensions.

Ideas involving branes and M-theory pushed aside, to some extent, the notion of single dimension strings, and the former ideas refashioned strings as being a limiting expression of a more expansive set of dimensional possibilities. Leaving M-theory aside for the moment, the notion of branes gives expression to a set of assumptions, methods, techniques, relationships, functions, and operations that explore various ways of representing various aspects of a system's dynamics through the descriptive potential of different facets of dimensionality [including points (a zero dimension brane) and strings (a one-dimensional brane), as well as multi-dimensional branes as well).

Such dimensionality could be construed in ways that involved various physical properties such as mass, charge, tension, force, and various other quantum properties, and, as well, dimensionality could involve the capacity to propagate through space in accordance with the rules of quantum mechanics. Strings and branes could interact with one another, and, in addition, branes could interact with other branes within contexts of higher dimensionality.

The concept of branes began attracting more attention after Joel Polchinski, Jin Dai, Rob Leigh, and, independently, Petr Horava, realized that a certain kind of brane – known as a D-brane – was present in the equations of string theory. Unlike closed strings, open strings have ends, and those open ends are required to terminate on D-branes.

Branes establish constraints on strings. However, strings also establish constraints on branes since strings determine the properties and dimensionality of the branes that also are setting limits on the associated strings.

Branes can extend into a multiplicity of dimensions. However, string theory permits branes to extend into no more than nine dimensions.

Branes, like strings, are a hypothetical concept. As such, branes, like strings, might, or might not, actually exist.

However, neither branes nor strings have to exist in any ontological sense for them to be able to have heuristic value. For example, reality might have dynamic features that are conducive to being descriptively and predictively engaged in fruitful ways through the language of branes and strings without requiring one to suppose that reality can be reduced down to such linguistic treatments, any more than philosophical treatments of existence require one to make reality a function of that kind of linguistic processing despite whatever value might come through such philosophical activity.

Branes are a way of organizing a set of operational properties and principles as a descriptive/predictive framework through which to engage phenomenal dynamics. One can build whatever properties one likes into them, and, then, proceed on to figure out how those properties might unfold over time as a function of such dimensional dynamics.

In short, branes encompass a systematic means through which to represent or refer to certain aspects of reality. This can be done for a variety of reasons including the desire to find new ways of solving or addressing on-going problems in physics and cosmology.

The theory of branes helped to make sense of certain particle phenomena that arose in the context of string theory that didn't appear to be a function of strings in an of themselves, and, therefore, brane theory alluded to the need for something more inclusive … that is, branes. Furthermore, the development of brane theory has led to the discovery of mathematical connections that seem to point in the direction of the need for a still deeper, more complex, and nuanced understanding that might, or might, not come in the form of M-theory.

However, in many ways, the current status of M-theory is as vague as is the meaning of the "M" in M-theory. That 'M' has been interpreted by different people to stand for: 'membrane', 'mysterious', 'magical', 'mystical', and, possibly, even an inverted version of the beginning letter of the last name of its originator: Edward Whitten.

Initially, theorists were somewhat cautious about including the idea of branes in models concerning the nature of reality. On the one hand, branes didn't require physics to operate in the same way at all points of space, and, on the other hand, branes didn't require dimensions to be all the same … branes permitted distinctions to be made in relation to whether dimensions extended off from the brane or whether those dimensions ran along the brane.

Both of the foregoing properties were problematic for theorists trying to model reality. Each of those properties of branes violated ideas about symmetry that were integral to the way many theorists thought about such things.

In 1995 Joel Polchinski once again upped the ante by demonstrating – at least, mathematically – that branes appeared to have an essential role to play in the future of string theory. Among other things, he showed that branes have the capacity to interact (via the properties of charge), as well as exhibit sensitivity to the presence of forces (through the property of tension).

In short, Polchinski had shown that branes could be treated as dynamical objects. They could act on and be acted upon by other objects, and they had a life that was, to some extent, quite independent of strings.

Another type of brane – seemingly -- is referred to as a p-brane. P-branes are associated with various solutions for the equations of general relativity and Andy Strominger, who had been studying p-branes, discovered that they might be able to account for the generation of some particles that fell beyond the – then -- current horizons of string theory, and, therefore, string theory might not account for the origins of all manner of particles.

Initially, the impression seemed to be that D-branes and p-branes were different kinds of entities. However, Joel Polchinski came to understand that at the energies where the predictions of general relativity and string theory tended to coincide, D-branes and p-branes were, more or less, one and the same thing.

In 1995, Ed Witten introduced another deeper, more startling form of duality [i.e., the property of two theories being the same despite assuming (or appearing to) different forms of description]. Essentially, Witten showed that, at low energies, there was one edition of ten dimensional superstring theory exhibiting the property of strong coupling (the degree to which strings interact) that was equivalent to an eleven dimensional version of supergravity (the theory that gives expression to a framework that exhibits Supersymmetry and contains gravity.)

One of the breakthroughs that came through the foregoing discovery of equivalency was the realization that perturbation theory (a mathematical technique for generating incrementally closer approximations to correct answers based on a problem one already knows how to solve) could now be used in conjunction with the eleven dimension theory of supergravity to solve problems that had been intractable in ten dimensional superstring theories exhibiting strong coupling. Perturbation theory did not work well in contexts involving strong interactions, but it did work well in contexts that involved weak interactions ... something that was present in eleven-dimensional theories of supergravity, and since Witten had shown that a ten-dimensional version of superstring theory was equivalent to an eleven-dimension version of supergravity, one could now use perturbation theory in conjunction with the latter framework to solve problems that had been intractable in ten-dimensional forms of superstrings exhibiting the property of strong coupling.

Witten further demonstrated that what previously had been considered to be five different editions of superstring theory were, actually, all equivalent to one another as well as being equivalent to eleven-dimensional supergravity theory. In short, all five superstring theories exhibited the property of duality ... they were all different but equivalent ways of describing the same underlying theory (somewhat similar to the way in which Heisenberg's matrix theory, Schrödinger's wave equation, Dirac's formulation, and Feynman's sum-over histories method are considered to be equivalent representations of quantum dynamics despite their differences).

Based on the foregoing considerations, Witten conjectured there must be some kind of all-encompassing single theory that could not only accommodate both ten-dimensional superstring theory as well as eleven-dimensional supergravity theory -- and do so irrespective of whether one was talking about weak interactions or strong interactions – but, as well, the theory that Witten was envisioning would be capable of developing and extending those theories to their full potential. He referred to this unknown, theoretical superstructure as M-theory.

To consider theories involving a different number of dimensions as being equivalent to one another has the surface appearance of belonging to the same category of magic as squaring the circle. This initial problem is further complicated when one understands that theories involving ten-dimensional superstrings do – as the term indicates -- contain strings, but theories involving eleven-dimensional supergravity do not contain any strings.

The seeming chasm between the two foregoing kinds of theories was bridged through the use of branes. More specifically, if one takes the extra dimension of supergravity and rolls it up into a very small circle, the 2-dimensiional brane that surrounds the compactified space looks like a string, and, as a result, one simultaneously cancels out the extra dimension while introducing strings into a framework that previously had been devoid of them.

While the dualities that were discovered among five, seemingly disparate theories of strings simplified things theoretically – that is, from the mathematical side of things -- the task of demonstrating how the Standard Model of physics arises naturally out of string theory became far more complicated. A multiplicity of different kinds of branes were conceivable -- with different sets of quantum particles, properties, forces, constants, and energies – and such branes could be organized in different ways among varying contexts of higher dimensionality.

In short, there were many, many possible worlds that could be constructed by stringing together, so to speak, different sets of possible variables involving branes and strings. Which, if any, of those combinations, or braneworlds, were capable of reflecting the Standard Model, or, which, if any, of those possible combinations were capable of providing a means through which to derive the Standard Model was unknown.

For a number of decades, string theorists have been putting together complex, Lego-like kits ... complete with a rulebook that specified the parameters through which construction might proceed. Those kits contained packets of different branes with different shapes and properties that could be assembled in a multiplicity of ways within a multiplicity of dimensional contexts.

The challenge was to use those kits (singly or in conjunction with other such kits) to construct a facsimile of the Standard Model in physics. Unfortunately, no one knew how to successfully meet the foregoing challenge ... not even those individuals who were responsible for inventing those kits knew how to proceed in a way that would be able to achieve such an end result.

They could assemble interesting, intriguing, and thought-provoking braneworlds using those kits. However, even though everyone knew all of the features (e.g., forces, particles, dynamics, energies, constants, properties, and so on) that needed to be incorporated into their braneworld-models to have a chance of being able to reflect the characteristics of the Standard model), they couldn't seem to find a way to successfully build those features into their models.

One also can make the foregoing point by approaching things from a slightly different perspective. More specifically, strings, as well as their successors – such as branes and M-theory – are, in some respects, comparable to the creation of the Klingon language. A set of means have been established – both in string theory/brane theory as well as in the Klingon language -- through which to describe, organize, analyze, critique, refer, allude, and model various realms of experience. However, in neither case, is one necessarily certain that the judgments that are being expressed through such systems of understanding are actually true or correct.

Understanding how a language/modeling system works or how it can be used/applied to descriptively engage reality is one thing. Understanding to what extent such a language or modeling system reflects the truth of things or is correct in its manner(s) of parsing reality is quite another matter.

For nearly three decades, thousands of qualified individuals, spread among several generations of scientists and mathematicians, have rigorously explored, minutely analyzed, and imaginatively expanded the horizons of multi-dimensional frameworks. Yet, despite all of the foregoing work, along with associated publications, there have been almost no experimental predictions generated through that work that can be tested ... with the possible exception of the vacuum energy issue with respect to which such theories have made predictions that are quite wrong.

Superstring theory – and its related, dualistic counterparts – is a highly speculative enterprise. While all manner of mathematical breakthroughs have been made in conjunction with the running of that enterprise, what, if anything, such speculation or associated breakthroughs have to do with concrete reality is still very much a mystery.

Given that M-theory continues to remain elusive and highly vague, multi-dimensional research involving superstrings, branes, and other related topics is pretty much an area of research that is still in search of its own foundations. Such research seems to be suspended between an unknown foundation and an unknown potential.

In light of the fact that string theorists and their brane relatives have, in a sense (as far as the real world is concerned), been stuck in neutral for decades (even as they add new pieces to the aforementioned kits, as well as introduce new rules and recipes to updated editions of the instruction books accompanying the kits in order to reflect discoveries that have been made), one might entertain the possibility that there is a need to take a critical step away from the usual way of understanding, filtering, and framing things via string theory and braneworlds and, instead, attempt to determine if there might be other possibilities that could bear more theoretical and empirical fruit than currently is the case with respect to strings, branes, and M-theory.

What follows is just a very preliminary and speculative sort of exercise. However, given that string theory and the notion of braneworlds seem to be having such difficulty in gaining traction with respect to inducing those ideas to work in the context of real world issues and problems, then what, really, it there to lose?

At the very worst, even if the following considerations are completely off-track, one will be no further behind than already seems to be the case in relation to string theory and braneworlds. And, in a somewhat more optimistic vein, maybe, there will be some ideas present in the following discussion that could induce someone to re-conceptualize things in a way that has more heuristic potential as far as trying to engage the reality problem is concerned in the context of the conditions of Final Jeopardy ... that is, being forced by the nature of the existential conditions in which we are immersed (especially with respect to having only a relatively limited time ... one life-span) to make judgments about the nature of reality before all the evidence is in (i.e., life comes to an end and proof continues to remain elusive).

From the perspective of string theory -- at least in terms of how some theorists came to understand things beginning in 1985 – one can accept the idea that open-ended strings are required to end on something (e.g., D-branes). Nonetheless, if strings actually exist in nature, then, the theoretical rule requiring strings to end on something – such as a brane -- seems somewhat arbitrary ... possibly more reflective of the requirements of the theory than the necessities of reality.

Indeed, this issue of arbitrariness is one of the problems associated with string theory in any of its versions since all editions of string theory have difficulty making contact with reality in any meaningful sense. For the most part, they all seem to be self-absorbed in the intricacies of their own modes of describing, thinking and understanding possibilities quite independently of whether, or not, any of those intricacies have something of importance to say about the nature of reality.

What if strings were <u>neither</u> a fundamental unit of nature, nor a particular kind of brane (i.e., one-dimensional)? What if strings arose as a function of the dynamics of dimensional interactions among both spatial and non-spatial dimensions?

Dimensionality should not be confused with the degrees of freedom inherent in the dimension of space. Dimensionality gives expression to that through which different kinds of potentiality are expressed (including, but not restricted, to the phenomenon of space).

Time is one kind of dimensional potentiality. Space is another kind of dimensional potentiality. Consciousness is, perhaps, another kind of dimensional potentiality. Intelligence might be a further mode of dimensional potentiality.

There are many different kinds of dimensionality possibilities. We intuit the presence of various kinds of dimensionality through the differential traces they leave behind in the detectors of the body, mind, and heart.

For the time being, I will consider the foregoing sorts of dimensional phenomena as givens. Just as physicists have long taken time and space as dimensional givens without trying to account for why and how such dimensions are possible, so too, there seem to be other kinds of dimensions that are present which cannot necessarily be reduced to being strict functions of either time or space but, nonetheless, can enter into relationships with both spatial and temporal dimensions.

Phenomena (whether observable or unobservable to human beings) give expression to the interaction of an array of dimensions. Time and space are just two of those dimensions.

String theorist, brane theorists and M-theorists have pursued things with the idea that nine of ten dimensions in superstring theory and ten of eleven dimensions in supergravity theory give expression to spatial dimensions. However, there is a difference between <u>representing</u> a dimension in spatial terms and claiming that the dimension being represented in that manner is necessarily spatial in character.

For purposes of quantification, calculation, and hoped-for tractability, some people are inclined to give conceptual representation to non-spatial possibilities through geometric modalities. This is done, for example, in physics with the dimension of time and also is done in many areas of science by limiting the idea of dimensionality to spatial degrees of freedom of one kind or another.

However, as indicated previously, one must be careful not to confuse the modality of representation with that which is being represented through such a modality. This is especially the case when, for the purposes of visualization and quantification, one attempts to translate a non-spatial metric into spatial terms.

In any event, perhaps actual strings – if they exist -- don't necessarily terminate on, or end on, a dimensional brane that is construed in spatial terms, as much as strings might display certain remnants of the potential of different, non-spatial dimensions to help give expression to strings (as a manifested, emergent phenomenon) in the context of dimensional interactions. However, in order for the foregoing phenomenal manifestation to be able to occur, the dimensions involved in the flow-through, interactional process must – to varying degrees -- be sensitive to the presence of whatever shaping potentials are present in other dimensions that are participating in the dynamic.

Dimensions flow through one another but not necessarily in a spatial sense. Spatial dimensions are one of the cloud chambers through which dimensional dynamics, in the form of physical events, make their presence known.

| Quantum Queries |

165

The mind is another such dimensional cloud chamber, and the heart -- in the sense of having the capacity to detect the presence of emotional, moral, and spiritual dimensional interactions – is a third kind of dimensional cloud chamber. Furthermore, there could be still other kinds of detectors that are possible as well.

Perhaps, as far as the actual nature of reality is concerned, strings – if they exist -- don't end on branes. Maybe strings could be conceived of as that which helps gives expression to the presence of dimensional dynamics that leave traces of those dynamics in the form of a set of properties that are manifested (e.g., in the form of strings) in that which has the capacity to detect the presence of such dynamics (or, at least, the outer most behavioral expression of those dynamics) such as occurs in the body, mind, and/or heart.

For example, a string that constitutes a phenomenal event (such as, in its simplest form, a particle being manifested In a given state) doesn't end on a 0-, 1-, 2-, or 3-brane. Instead, time flows through spatial dimensions and leaves a trace of its influence behind that helps shape the character of an event that occurs in conjunction with certain spatial co-ordinates, along with whatever other properties (generated through the presence of other dimensional dynamics) are associated with the manifestation of a given string event. Moreover, other non-spatial dimensions – such as consciousness, intelligence, understanding, imagination, etc -- might be involved in such a shaping/detecting process, and the manifested event that is associated with those co-ordinates is a string -- a sign, marker, or indication -- that two, or more, dimensions have flowed through one another and left traces of themselves that collectively given expression to an event of a given character or set of properties contributed by the different dimensions and detected by that (e.g., the mind and body) which has the capacity, within limits to detect certain facets of those dynamics.

An issue that tends to disappear when dimensionality is considered as being something other than spatial in character is the notion of compactification. Compactification is the process of curling things up in various kinds of compact spaces (e.g., Calabi-Yau manifolds) to help explain why we only experience a four-dimensional world even as those extra, compactified dimensions play a role in shaping the way in which phenomena are manifested through those four dimensions.

However, possibly, the reason why we experience phenomena in the apparent context of four dimensions is because like time, the other dimensions, are largely invisible in their manner of being present. We infer the presence of time because, among other things, physical phenomena change, and, as a result, time would seem to be one of the necessary dimensional components that need to be present in order for that kind of change to be able to occur.

We don't see time directly. But we do experience its presence.

Actually, in the same way, one also might argue that we don't see space directly. Rather, we experience and infer its presence through the degrees of freedom it provides for movement and dynamics.

As such, space is not multi-dimensional. Space constitutes just one dimension with an unknown number of degrees of freedom.

We know there are at least three degrees of freedom inherent in the spatial dimension because we experience those degrees of freedom in everyday life. Whether, or not, there are more than three degrees of freedom inherent in space is a speculative exercise that is pursued by various mathematicians, physicists, cosmologists, psychologists, and philosophers.

When someone claims there are nine or ten spatial dimensions involved in this or that phenomenon, at least two very different kinds of things could be meant. On the one hand, a individual might mean that the phenomenon in question really is, in some sense, a function of ten separate spatial dimensions interacting with one another such that each, distinct, spatial dimension gives expression to certain kinds of shaping potential with respect to the overall structural character of the phenomenon being manifested through the collective contributions of the potential inherent in those spatial dimensions. Or, on the other hand, the aforementioned nine or ten spatial dimensions are intended to refer to a complex, descriptive co-ordinate system consisting of a separate axis (beyond 'x', 'y', and 'z') for each variable that is being represented by a given spatial dimension (i.e., an n-tuple), but such dimensionality is only representational in the context of such a co-ordinate system and that sort of descriptive dimensionality is not existential or ontological in nature beyond its existence as a system for modeling various aspects of reality.

The two foregoing meanings are often conflated with one another, and, in the process, considerable confusion can arise. Compactification is rooted in an understanding that stipulates spatial dimensions are real and, therefore, their observed absence in the world of everyday experience must be explained away by invoking the process of compactification (although no explanation is ever given of what is responsible for the process of becoming compactified or how that process is accomplished). However, if the references to extra spatial dimensionality are purely representational (that is, it constitutes a way of arranging variables in a complex, co-ordinate system in order to try to gauge the functional relationships among those variables), then, there is not necessarily any need to introduce the notion of compactification.

Similarly, in an earlier chapter, I mentioned the possibility that quantum objects are not necessarily both waves and particles (i.e., wavicles), but, instead, they might be neither. If dimensionality is considered to be basic, then, wave and particle phenomena are not a function of one and the same thing behaving differently under different conditions but, instead, those phenomena could give expression to the way an array of ontological (not spatial) dimensions differentially interact with one another from one set of circumstances to the next.

The foregoing issues will be discussed further – at least, to some extent -- in later volumes of this book. The purpose of the foregoing exercise is to suggest some possibilities about how one might go about beginning to re-conceptualize the notion of dimensionality in order to begin to move in a hermeneutical direction that is somewhat different from the highly spatialized framework of string theory ... that is, in string theory (including its updated version of branes) dimensions tend to be represented through the imagery of, and considered in terms of, being a spatialized phenomenon in which each degree of freedom of space is considered to be a different dimension rather than that the dimension of space is being parsed in different ways in the context of an attempt to represent the dynamics of phenomena.

In the previous chapter a criticism was directed toward a journal editor who referred to Peter Higgs initial article on the relationship between the origins of mass and the process of spontaneous symmetry breaking – which, with a couple of paragraphs added, was the basis of his being awarded a Nobel Prize -- as "having no relevance to particle physics." Of course, in accordance with the principles of 20/20 hindsight, and as indicated in the earlier chapter, the editor being referenced here obviously didn't know as much about physics as he thought, but, in fairness, what is obvious after the fact is not always so clear before the fact.

There can be mistakes of omission, as well as mistakes of commission, when it comes to what is, and is not, published in journals of science. The aforementioned Higgs affair was, by and large, a mistake of omission ... something should have been published but was not.

There also have been errors of commission that have taken place with respect to journal articles. These are articles that did not meet some minimal level of scientific rigor, and, yet, were published nonetheless.

| Quantum Queries |

Sometimes – and, this can be so for a variety of reasons, both good and bad -- a benefit of the doubt is given to the putative value of someone's ideas when that person might not necessarily deserve that sort of consideration. For instance, in 1996, Alan Sokal – a mathematical physicist -- wrote an article entitled: "Transgressing the Boundaries: Toward a Transformative Hermeneutics of Quantum Gravity."

The article was submitted to *Social Text*, a prestigious, academic journal in postmodern cultural studies. The article was accepted for publication.

Among other things, Sokal's piece argued that quantum gravity was a linguistic and social construct. Sokal couched his argument in vague, relatively meaningless, and nonsensical terms but did so a way that, on the surface at least, appeared to create a intelligible line of argument concerning an issue of some importance to science and was doing so from a postmodern perspective.

Purportedly, Sokal did what he did to test the degree of intellectual rigor involved in the process through which humanistic approaches to issues of science were vetted. The fact his article was accepted demonstrated – at least in this case – that such a vetting process seemed to lack a great deal of intellectual integrity.

Some people looked at the Sokal hoax as demonstrating the superiority of the way things were done in the sciences as opposed to the way things were done in the humanities. However, such a conclusion might have to be modified, to some degree, in light of the following discussion.

More specifically, shortly after the turn of the century (2000), two brothers in France – Igor and Grichka Bogdanov – had written several theses (for which they were awarded doctorates) -- and a number of journal articles that were referenced in at least one of the foregoing doctorates. The latter articles had been published in a number of prestigious journals

– including *Annals of Physics* as well as *Classical and Quantum Gravity*, and, also, in several other lesser-known journals.

Apparently, numerous editors and referees had read through the articles and considered them acceptable for publication. Unfortunately, none of the foregoing editors or referees appeared to have taken their responsibilities very seriously ... at least, that is one of the possible conclusions drawn by Dr. Peter Woit, a mathematical physicist.

Dr. Woit's curiosity had been tweaked by the Bogdanov brothers because he had received a number of e-mails indicating that some sort of reverse-Sokal hoax might have been perpetrated by the aforementioned two brothers, but this time, the affair carried potentially unsettling ramifications for the intellectual rigor associated with a number of science journals rather than a single journal in the humanities. Dr. Woit went through the forgoing articles and discovered that, like the Sokal hoax, mentioned earlier, the articles were full of nonsense.

Woit indicates that at the time he became interested in the foregoing issue he was informed that individuals in the Harvard string group apparently couldn't determine – at least, based on a superficial reading of the material – whether, or not, the articles were fraudulent. When such individuals were informed, at one point, that the papers were being treated as fraudulent work, some of the individuals in the Harvard string group were joking about how obvious the fraudulent character of that material was, but, when the members of the Harvard string group later were informed that the authors of the papers were real professors in physics and that the work was being considered to be credible, the aforementioned individuals backed off from their previous position and entertained the possibility that maybe those materials were, after all, credible and had scientific value.

Senior personnel from the journals that had permitted the articles to be published later admitted that mistakes had been made with respect to letting those materials be published. Yet, an after-the-fact admission of culpability doesn't necessarily automatically cancel out a before-the fact agreement of editors and referees that such articles constituted acceptable scientific work, and this tends to make one wonder about the integrity of such a process.

Moreover, as embarrassing as the *Social Text*/Alan Sokal hoax was for the humanities, the Bogdanov brothers had induced five sets of editors and referees at different scientific journals, plus one dissertation committee consisting of scientists and mathematicians (in the case of Igor Bogdanov) to treat their writings as being scientifically reputable. Seemingly, the Bogdanov affair involving scientists and scientific journals was of a far more egregious nature than the *Social Text* hoax involving the humanities.

Were the foregoing five sets of editors and referees just too lazy to do their jobs? Or, were they – like the editor who rejected Higgs's work – dealing with something that they did not understand but passed judgment on anyway (negatively in the case of the Higgs rejection, and positively in the case of the Bogdanov brothers) rather than admit that they didn't understand what was being said in either instance.

Similarly, there is a great deal of work that has been going on for more than three decades with respect to strings, superstrings, branes, and related subjects that is very difficult to assess as to its probative value with respect to being able to successfully delineate – in the near-future, if ever -- the nature of reality. And, yet, many degrees, postdoctoral fellowships, research positions, professorships, and grants are being awarded, as well as articles and books being written, in all of the foregoing areas on the basis of what might only be scientific/mathematical version of the 'Emperor's Clothes' ... a process that is stringing us all along – including the theorists – and moving us down what might prove to be a theoretical cul-de-sac.

Eventually, the string/brane/M-theorists might still win the day and have their ideas vindicated and shown to be relevant with respect to solving the reality problem. Right now, this is not the case, and those of us who are interested in dealing with the issue of the reality problem in the context of the Final Jeopardy issue might just have to move on since time is running out and continuing to invest time in brane-futures seems far too risky.

| Quantum Queries |

Chapter 6: Quantum Illusions

Perhaps no aspect of quantum mechanics has been cited by non-physicists more frequently – and with, possibly, less understanding -- than has Schrödinger's famous thought experiment involving a cat. A variety of individuals seem to believe that Schrödinger's thought experiment constitutes proof that the Copenhagen theory of quantum interpretation is correct, when nothing could be further from the truth.

Schrödinger introduced his hapless cat in a 1935 article. In that paper, Schrödinger asked his readers to imagine the following possibility.

A cat has been placed in a steel chamber of some kind. Within that same steel chamber – in an appropriately cat-proofed sub-section of the interior portion of that enclosed structure – is an apparatus consisting of various components.

Among these components is a Geiger counter situated near a tiny bit of radioactive material. The radioactive properties of that material are such that the likelihood of one of its atoms decaying during the next hour is very high, but not certain.

The Geiger counter is hooked to a hammer in such a way that if a click is registered due to the detection of radioactive decay, a hammer will be released from a latch and, then, will fall toward a glass flask. If the hammer falls, it will fall with a force that is capable of breaking a small flask that contains hydrocyanic acid.

If, as a result of the foregoing scenario, the flask breaks at some point during the hour in which the cat is kept in the steel chamber, the substance in the flask will be released, and the cat will die of hydrocyanic poisoning. If the flask does not break, then, the cat will not die from that kind of poisoning … although -- depending on how small and airtight the steel chamber is -- the cat might have died from asphyxiation due to the build up of CO_2 and the disappearance of oxygen during the hour-long period in which the experiment is being run.

Assuming that the steel chamber is of a reasonable size with a sufficient source of oxygen and an adequate means of dissipating the CO_2 being exhaled by the cat, what is the ontological state of the cat during the hour in which the cat occupies the closed steel chamber? There are several possible ways to respond to the foregoing question.

The principle of superposition that is referenced by the Copenhagen theory of quantum interpretation holds that when calculations are made concerning the ontological status of the cat within the experimental system, all possible outcomes exist simultaneously. In other words, the Ψ (Psi) function (whose permitted values are determined through Schrödinger's wave equation) that is used to represent the foregoing experimental set-up would describe the cat as being smeared out over a probability distribution in accordance with the principle of superposition.

In other words, the probability distribution representing the state of the cat within the closed chamber indicates that the cat's possible states involve: (a) being alive; (b) being dead; and, (c) in the process of dying, and, therefore, one must include all three possibilities in any description of the cat's state while in the steel chamber. As such, the cat is described as being simultaneously living, dead _and_ dying.

There is another possible response to the previous question concerning what the ontological status of a cat will be while in the aforementioned experimental set-up and prior to the steel chamber being opened. Either the cat is alive, dead, or in the process of dying, but the cat is not in all three states at once.

Schrödinger would like to know the answer to the following questions: If the Copenhagen theory of quantum interpretation is correct, when -- and how – does the actual outcome (the one that is seen when the steel chamber is opened) become realized from the set of probabilities that initially described all possible outcomes in a given system as existing simultaneously? Stated in an alternative fashion: How does a concrete, empirical

| Quantum Queries |

outcome arise out of an indeterminate, probabilistic starting state?

The Copenhagen school of quantum interpretation maintains that a definite outcome arises out of the aforementioned amalgamation of states that collectively have been given expression through the superposition principle by means of the measurement process (and there are other theoretical interpretations involving the Schrödinger cat experiment, some of which will be discussed later in this book). As long as such a measurement is not made, then, allegedly, the ontological condition of the system is described by the way Schrödinger's wave equation establishes the permitted Ψ (Psi) function for that system, and this information can be used to generate a probability distribution that includes all possibilities for such a system ... possibilities that – according to the Copenhagen theory of quantum interpretation -- continue to exist simultaneously until measurement decides the matter.

Schrödinger invented a wave equation that helped calculate the values of the wave function Ψ (Psi) that is associated with a given system. However, Schrödinger was not happy with the way his equation had been hijacked by the Copenhagen theory of quantum interpretation and turned into an ontological perspective rather than merely being used as a methodological tool.

Schrödinger hoped that his thought experiment involving the cat would serve as a way of lending support to Einstein's position concerning the underlying – although, possibly, currently unknown -- nature of reality by not only pointing out how ludicrous Schrödinger believed the Copenhagen perspective sounded but, as well, by showing that there were important questions (noted previously) that the Copenhagen interpretation of quantum theory could not answer. Schrödinger, like Einstein, believed that reality is more that some sort of vague realm of inherently indeterminate and probabilistic entities as proponents for the Copenhagen theory of quantum interpretation – led by Bohr – appeared to be claiming

The principle of superposition is the hypothetical basis underlying a particular mode of describing and interpreting quantum phenomena rather than constituting a law that necessarily reflects the structural character of reality. Although the phenomenon of interference (especially in the case of double-slit experiments) is often cited as evidence indicating that the principle of superposition is operating on the quantum level, one might want to separate the data being used to point in the direction of the phenomenon of interference from the hermeneutical system that is being used to interpret that data and phenomenon.

More specifically, while the principle of superposition does offer a way to <u>interpret</u> the pattern on the screen that arises from, say, a double-slit experiment, one need not necessarily suppose that the screen exhibits the pattern it does because, <u>ontologically</u>, electrons or photons simultaneously occupy all possible quantum states (and, therefore, all possible routes) while in route to the screen. No one knows what is taking place when two streams of photons or electrons are being run through a double slit device that, subsequently, generates data on a screen in the form of a pattern which suggests not only that interference of some kind has taken place but, as well, that all of the possible pathways to the screen appear to be represented in that pattern.

In ontological, dynamic terms, what is entailed by the idea that a particle simultaneously follows all possible pathways to a screen (pathways whose number could, theoretically, be infinite in number)? How does a particle run through an infinite number of paths in not only finite time but in accordance with a very short subset of finite time (the velocity of these particles is fairly fast, and the distance to the screen is not great)?

Even if the number of possible pathways were not infinite in number, one would like to know how the dynamic works ontologically (and not just theoretically)? How does a particle 'know' which paths are possible? That is, how does a particle go about exploring all possible paths – even if finite in number -- simultaneously?

The absence of any plausible explanations concerning the foregoing questions is problematic. If one doesn't know how something actually works, then, why should one suppose that any given quantum entity actually explores all possible pathways simultaneously, or, at least, does so within the time interval that transpires as that particle passes through the double-slit on its way to the photographic screen?

Is it possible that one, or another, modality of the phenomenon of entanglement is taking place as quantum entities go through the slits? After all, if entanglement of some sort can occur with quantum entities that are distant from one another, then, why isn't it possible that an entanglement phenomenon of some kind might be involved with entities that are separated by only relatively small distances … especially given that we really don't know all that much about what actually takes place during the process of entanglement (that is, we know that it occurs, but we do not how it takes place, or to what extent it occurs, or whether, or not, there are different kinds of entanglement)?

Of course, across the very tiny time scales that mark the journey of particles from an energy source to a detection screen via way of a double-slit device, all of the possibilities inherent in a set of electrons or photons might show up at different intervals and points that help shape the pattern on the screen. Nonetheless, this doesn't necessarily mean that all such possibilities are manifesting themselves simultaneously as required by the principle of superposition.

Moreover, even if it were true that the principle of superposition accurately describes what happens in some cases on the quantum level, there is no evidence demonstrating that such a principle also carries over to, and is operative on, a non-quantum or macro level. As far as I know, no one has shown that the principle of superposition (even if it accurately describes what takes place in the quantum world) is scale invariant with respect to all ontological levels ... from the smallest, to the largest (including the level on which Schrödinger's cat experiment would take place).

In addition, as far as I know, there is an absence of evidence indicating that reality consists of nothing more than a set of probability distributions in which all possibilities described through those distributions exist simultaneously. Instead, such a conceptual perspective seems to be imposed on reality irrespective of what the actual character of reality might be.

One of the weaknesses underlying the ontological claims associated with the principle of superposition is that there is no coherent account of: (a) How measurement is capable of engaging the possibilities that allegedly exist simultaneously and, then, (b) how that process of measurement selects the portion of such a probability distribution that will turn indeterminacy into determinacy. In fact, the foregoing procedure does not really constitute a process of measurement in any traditional sense (analyzing a sample in terms of a calibrated standard) but, rather, under the circumstances of superposition, measurement becomes a process for inducing a particular portion of the probability distribution to manifest itself ... and, even then, one is uncertain what the precise nature of that induction process is.

A value has been identified through what is nominally being called a "measurement". While the "measurement" is an outcome of sorts, there is no defensible reason that can be given for why one value has emerged through such a process rather than some other value ... other than that it is consistent with the parameters of probability within which such values occur.

Ignorance concerning the nature of a system – e.g., the ontological status of the cat in the steel chamber during the running time of the experiment – is being described in a way (i.e., through the principle of superposition) that seeks to claim that one's epistemological status in relation to the cat experiment is not really a matter of relative ignorance but constitutes an epistemological state in which one knows that the cat is alive, dead, and in the process of dying. However, one knows nothing of the sort.

Rather, one knows there are three possible states for the cat to be in: The cat is alive, dead, or in the process of dying. Nonetheless, nothing in the foregoing condition of understanding can demonstrate that all three of those possibilities are simultaneously true or that any one of them is true ... only that all three of them are possible and that they seem to exhaust the possibilities associated with such an experimental set-up.

Invoking the principle of superposition in the case of Schrödinger's cat is to make an epistemological claim that cannot be verified. Through a process of magical thinking, the principle of superposition seeks to leverage the associated hermeneutical issue of indeterminacy and invest it with an ontological condition that is not logically or empirically tenable.

In short, there is an illegitimate conflation that is taking place between, on the one hand, the principle of superposition, and, on the other hand, the Ψ (Psi) function's permitted values are determined through Schrödinger's equation. The principle of superposition purports to be a viable interpretation of the ontological significance of the probability distribution that is generated through Schrödinger's equation, and, yet, there is nothing to support such a declaration except the claim itself.

The principle of superposition is rooted in a failure to distinguish between a description of something and the actual character of whatever is being described. An abstract description involving probabilities appears to be reified into a concrete set of simultaneously existing realities.

According to Jean Piaget, a Swiss developmental psychologist, infants acquire an understanding concerning the permanence of objects near the end of the sensorimotor stage of development. Object permanence has to do with the ability of an infant to be able to show evidence that he or she understands that when a once visible object (say a ball) is hidden behind (for example, a pillow) or under something (e.g., a blanket), the infant, nonetheless, knows that the object still exists but is now in a hidden or cloaked condition.

The sensorimotor stage of development refers to a period lasting from birth until approximately two years of age. Prior to the latter part of that stage of development, Piaget believed that infants do not possess the concept of object permanence, and he generated a fair amount of experimental data in an attempt to support that belief.

However, more recent studies in developmental psychology have challenged Piaget's beliefs concerning when object permanence is acquired as well as what constitutes evidence that such a concept has, or has not, been acquired. Much recent research suggests that object permanence might be acquired more than a year and a half earlier than Piaget maintained … indeed, possibly as early as three months of age.

Irrespective of the timing issue – that is, irrespective of the point during development when an infant acquires the concept of object permanence – all research involving psychological development (at least, from Piaget onward) seems to indicate one thing. Human beings do have the capacity to understand that an object hasn't necessarily de-materialized simply because it is not visible.

Before being placed in the experimental steel chamber, Schrödinger's cat is concrete and visible. Once the chamber is closed, the principle of superposition claims that the cat transforms into a set of simultaneously existing probabilities … that is, the cat is no longer a concrete cat but, rather, it has become an abstract representation of a cat that is simultaneously existing in different quantum states, and such an array of possibilities can be described through an appropriately calculated probability distribution.

The Copenhagen interpretation of quantum theory cannot say how the foregoing transformation from a concrete visible object to a set of simultaneous possibilities takes place. The Copenhagen interpretation of quantum theory cannot say what happens to the energy that – prior to entry into the steel chamber – has been actively organized in keeping the visible cat alive or how that energy is transformed into something that maintains an array of probabilities rather than lived actualities.

Why should one suppose that an object – i.e., the cat – that is now hidden inside the steel chamber – is no longer a concrete cat? Why should one suppose that our understanding of object permanence is incorrect?

To be sure, there are circumstances involving magic (whether done on a stage or in close-up conditions) in which a person's expectations about object permanence can be thwarted through sleight-of-hand, misdirection, and the capacity of a magician to create illusions -- with or without the aid of props -- by managing an audience's perceptual and/or mental activities. But, there is no real magic involved in the foregoing tricks such that objects suddenly actually cease to exist or objects materialize out of nothingness.

The principle of superposition seems to say that our understanding of object permanence – on both the macro and micro level -- is wrong. That principle stipulates that what we take to be object permanence is really nothing but the capacity of special kinds of probability distributions to give expression to specific instances of a set of properties at a given point in time and under certain circumstances. (e.g., during the process of measurement).

Like all good illusionists, proponents of the Copenhagen school of quantum interpretation never explain how the trick is done that permits abstract probabilities to materialize in one concrete form rather than another. Instead, all one has to do is to tap on the experimental steel chamber three times with one's quantum wand, lift the lid of that chamber, and Dorothy is back in Kansas where cats (and dogs) appear to be just cats (and dogs).

The issue of object permanence touches on the issue of realism. Realism is predicated on the belief that whatever can be seen, touched, smelled, and so on at a given point in time, doesn't necessarily cease to exist just because it is hidden or out of sight ... although, of course, something could happen to that something when it is out of sight that does actually cause the object to cease to exist.

Given time and the right set of conditions, many objects do decay and cease to exist. So, there are limits that can be placed on what is meant by the notion of object permanence.

For instance, on the quantum level, there is experimental evidence indicating that neutrinos transform into different versions of themselves. In addition, from time to time, quarks can assume different flavors, and under the right circumstances, neutrons can transform into protons. Moreover, virtual particles seem to have the capacity to blink into and out of existence within the vacuum of space.

Nonetheless, all of the foregoing possibilities take place within the parameters that are a function of the interaction of forces, energies, fields, and structural properties of the particles that are present during such interactions. Even when quantum objects don't have prolonged permanence, per se, the dynamics that are reflected in transformations involving those objects still have a sense of permanence about them.

The concept of object permanence is, in its own way, an expression of the principle of conservation at work. Whether something exists – at least temporarily -- as a quantum object or exists as a transformation of such an object, there is an underlying reality that is present and doesn't disappear.

| Quantum Queries |

This is at the very heart of the idea of a law of nature. Science might capture various facets of the structural properties of the way in which nature operates, but that to which science alludes exists independently of the methods, theories, and interpretations that are used to try to descriptively represent various dimensions of reality.

Furthermore, within the short run – say the running time of one hour required by the Schrödinger cat experiment – most of us have accumulated very little experiential evidence, if any, that would lend credibility to the idea that once the lid to the steel chamber is closed, then the cat suddenly transforms into a set of probability distributions consisting of simultaneously existing quantum states, or that once the lid is raised, then, the probability distributions giving expression to simultaneously existing quantum states suddenly materialize in the form of a particular cat that is living, dead, or in the process of dying from hydrocyanic acid poisoning.

If the cat is dead when the lid is raised, then, how did such a state arise? Presumably, death by hydrocyanic acid is a fairly complex process involving a number of biochemical reactions, and, therefore, one wonders how the probability distributions for all of these biochemical quantum states arise, interact with one another, and lead to one result rather than another?

How does the probability distribution that constitutes the decaying of a radioactive material engage the probability distribution that gives expression to the simultaneously existing quantum states that constitute a latch holding a hammer in place, as well as a hammer falling, and, in addition, a glass vile being broken by a falling hammer? How does the probability distribution representing the release of simultaneously existing quantum states of hydrocyanic acid engage the probability distribution of simultaneously existing quantum states constituting a cat in order to bring about – possibly -- death?

Does death give expression to just one particular quantum state, or is death a process that takes time? From whence did such a quantum state or quantum states arise and how did it (they) acquire its (their) properties?

How does the experimental set-up 'know' that death is one of the possible outcomes of that set-up, and, as a result, is able to include that quantum state among the possibilities that make up the probability distribution giving expression to the quantum states that describe that experimental set-up? How does the cat 'know' to include the possibility of death by hydrocyanic acid poisoning among its quantum states when it doesn't necessarily even know what hydrocyanic acid is or whether, or not, it is present in the experimental chamber, or whether, or not, it can cause the cat to die?

What organizes the nature of reality in a way that can be reflected through probability distributions? Should one suppose that, such probabilities just spring into existence uncaused and for no reason whatsoever, and, if so, why should we make such a supposition?

Ontologically, what does it even mean for a simple, quantum entity (not a complex entity such as a cat) to be in a state of superposition? Supposedly, all of the possible states for such an entity are present, in some sense, as probabilities.

What does it mean to exist as a set of probabilities? Exist where? Exist how?

Is the probability distribution a basic or fundamental ontological reality of some kind? If so, what kind of reality is it, and how does it arise in the form it has?

What quantum values shape those probabilities, or do the probabilities shape the character of the quantum values that are present? In either case, what is the specific nature of the dynamics that give expression to such shaping processes?

What sustains that set of probabilities? What transforms it from a probability to an actual reality?

Quantum Queries

Everything appears to be assumed by the principle of superposition. Interpretation is reality, and, therefore, nothing seems to have to be proven.

Schrödinger's equation works – or, so, we are told by some individuals – because the principle of superposition governs the nature of reality ... that is, all possibilities simultaneously exist amidst an array of probability distributions that are particularized in the form of events by means of a still mysterious process known as measurement. Unfortunately, instead of admitting that no one knows why Schrödinger's equation works (only that it does), some people have confabulated an entire philosophy in an effort, apparently, to avoid having to look too deeply into the mysterious depths of reality or in order to avoid having to deal with the possibility that such bottomless depths might somehow be staring back and, maybe, even making note of the many ways in which we go about distorting its nature.

Adults who should know better – and who (theoretically) have been trained to be sensitive to the issues of critically examining issues and requiring empirical data to demonstrate the possible truth of something – have gobbled down an indigestible meal of philosophical vagueness and continue to sing the praises of such vagueness as if were meaningful in some essential sense despite the fact that such a perspective appears to be devoid of both critical thinking and supporting empirical data. Of course, at some point in the future, someone might be able to demonstrate that the world really does exist in all quantum states at once and that quantum entities really do travel to a given point in space through all possible routes simultaneously, but until that time, the principle of superposition is nothing more than an unproven hypothesis that is associated with a wave equation and a wave function that can be used to determine probable values without necessarily having to conclude that all probable values must exist simultaneously.

Existing simultaneously in the form of a potential is not necessarily the same thing as existing simultaneously in the form of an active reality. The wave function Ψ (Psi) gives expression to the potential of a system and that potential is parsed by the probability distribution that is used to make predictions concerning how such a potential might manifest itself or unfold under different conditions over time.

For example, depending on circumstances, a quantum entity might have the potential to be in a high-energy state or in a low energy state, but, ontologically speaking, it is not in both energy states at the same time. The structural character of that quantum entity is such that it has the capacity or potential to take on different values, and the identity of the values it takes on will depend on circumstances.

Thus, the way in which a given quantum entity is manifested will depend on two things. On the one hand, it depends on whatever the inherent structural character of that entity is, and, on the other hand, the manifested character will depend on the nature of the dynamic between that structure and the surrounding physical circumstances in which it exists.

The principle of superposition describes the potential of a system. That system consists of the structural character of a given quantum entity interacting with the contingent circumstances that are present within the boundaries of such a system.

A system in the foregoing sense gives expression to that which is described in classical mechanics by means of the Hamiltonian or Lagrangian (These are related but different ways of representing the total energy/dynamics of a system). The idea of superposition refers to the potential being described in conjunction with the quantum dynamics of such a system.

Interactional possibilities are descriptively represented through probabilities. Probabilities acquire their properties as a function of the dynamics generated through the foregoing interactions.

Probabilities are predicated on, and derived from, the existence of a dynamic whose details are unknown. The surface features or outcomes of that dynamic can be described through probabilities, but the underlying properties of that dynamic elude the grasp of the mathematical process through which those probabilities are calculated.

The Copenhagen theory of quantum interpretation claims that the probabilities entailed by the principle of superposition are a fundamental reality. There is nothing beneath them, and no other level of reality eludes those calculations … there are no hidden variables.

Unfortunately – and has been outlined, to some extent, over the last six or seven pages – there appears to be a relative dearth of data or arguments that accompany the foregoing interpretive perspective which are capable of demonstrating why that perspective should be considered to constitute a plausible account of the nature of reality. As it stands, that interpretive perspective appears to be nothing more than a methodological way of engaging reality that makes the latter (i.e., reality) a function of the former (i.e., methodology).

In other words, the principle of superposition is a way of understanding things that seems to fail to consider the possibility that methodology is just a way of filtering and framing experiential data that are derived from having engaged reality in certain ways. As such, that method does not necessarily constitute or completely reflect the nature of reality itself

To be sure, good methodology has the ability to highlight and characterize various properties that are made possible by the nature of reality as the latter is engaged through one or another kind of hermeneutical methodology. However, such methodology doesn't get to dictate what reality can and can't be, and this, essentially, is what the Copenhagen theory of quantum interpretation seems to be doing (The same kind of mistake was made in conjunction with interpreting the significance of Heisenberg's uncertainty principle.).

The principle of superposition does not give expression to the active, or manifest, state of a quantum entity. Rather, the principle of superposition gives descriptive – not ontological -- expression to the array of possibilities that are potential ways in which the underlying, unknown dynamic might manifest itself.

Measurement is a sampling process that engages the foregoing dynamic. Measurement does not engage quantum entities in a condition of superposition, but, rather, the principle of superposition alludes to a property of the method used to descriptively represent the potential of the dynamic that is being sampled through measurement.

During the mid-1950s, Hugh Everett III approached many of the foregoing issues from a very different theoretical perspective. His conceptual position gave expression to a detour – at least from the perspective of the Copenhagen theory of quantum interpretation -- in relation to, among other things, the issue of measurement as well as the principle of superposition.

As indicated earlier in this chapter, physicists employ the wave function to mathematically describe the states that are associated with a quantum entity. A wave function encompasses all of the possible quantum states for a given system, along with a set of probabilities that are assigned to an array of possibilities indicating the likelihood that any one of those quantum states will occur if a given particle is engaged through a process of measurement or if that particle is involved in an interaction of some kind under the conditions being described through such a wave function.

According to the Copenhagen theory of quantum interpretation, there are two broad levels of reality. These consist of the quantum realm and the classical realms (the latter domain gives expression to the macro world with which we are familiar through the experiences of everyday life ... including measurement.).

| Quantum Queries |

The quantum world operates in accordance with the principle of superposition – that is, the configurations described by the wave function exist simultaneously with one another. On the other hand, the classical world of measurement somehow (and this continues to be an unsolved problem for the Copenhagen theory of quantum interpretation) becomes correlated with just one of the foregoing quantum state possibilities, and, therefore, treats measured-reality as being a narrowed-down version of the original condition of superposition.

The classical measurement process lends an empirical meaning to phenomena. However, according to the Copenhagen perspective, that process does so at the cost of distorting the actual nature of quantum reality ... that is, the quantum realm's assumed natural condition of operating in accordance with the principle of superposition.

For example, Schrödinger's equation describes how the wave function will unfold over time. That description depicts the foregoing unfolding process as being continuous and deterministic.

Yet, when a measurement it taken, both the continuous and deterministic nature of that equation seem to fall by the wayside. Only one possibility from amongst a continuum of possibilities is given expression through a measurement, and, in addition, there seems to be no discernible basis for justifiably claiming that the value produced through such a measurement process constitutes a deterministic reflection of the possibilities inherent in the system being described by the Schrödinger equation ... instead, the value produced through the measurement process seems random or arbitrary.

Furthermore, the Copenhagen theory of quantum interpretation is unable to establish the character of the boundary conditions that demarcate where the quantum world leaves off and the classical world begins, and, as well, that theory is unable to explain how the two realms interact across such an alleged set of boundary conditions. However, the individual(s) performing the measurement are considered to be part of the classical world, whereas the entities operating in accordance with the principle of superposition are part of the quantum world.

From the perspective of the Copenhagen theory of quantum interpretation, a measurement collapses the wave function. In some unknown fashion, the measurement process breaches the boundary or barrier separating the classical and quantum worlds and generates a value ... the measurement.

Hugh Everett argues that there is no boundary separating the classical and quantum worlds. Moreover, measurements do not disrupt the manner in which the wave function unfolds over time.

According to Everett, the so-called classical world, as well as the quantum realm, each operates in accordance with the principle of superposition. Both levels are linked through one and the same wave function, and, in addition, the continuity and determinate nature of Schrödinger's equation is not violated by a measurement process, but, instead, such measurements merely induce the system being described by the unfolding wave function to bifurcate in a such a way that, eventually, all possibilities that can happen will happen.

Everett believes that the potential of a wave function that is governed by the principle of superposition will be realized over time. Measurement – and, in a sense, every observation is a measurement -- is the process of unlocking the potential of the possibilities inherent in the principle of superposition that is given expression through the wave function that describes the interaction of the observer and the observed.

On each occasion that a measurement or observation takes place, a possibility inherent in the potential of the wave function is realized in the form of an ontological splitting or bifurcation that constitutes a dimension of the original wave function's potential. Measurement and observation don't interrupt the continuity of Schrödinger's equation, but, instead, help that continuity to come to fruition in the real world through the process of splitting or bifurcation.

Every bifurcated expression of the potential of a given wave function constitutes its own self-contained realm. Differences among those self-contained realms reflect the manner in which a given instance of bifurcation has explored or realized one dimension of the potential of a wave function's property of superposition rather than explored or realized some other dimension of that potential.

In principle, Everett's perspective doesn't even require the presence of measurements or observations to generate the foregoing sorts of bifurcations. Every event, every dynamic, every interaction, and every happening brings about the process of bifurcation.

'In the beginning' was the wave function for the entire universe. And, it was good.

The history of the universe from that time forward is a matter of keeping track of the ensuing set of bifurcations. These splits give expression to the superposition of the quantum possibilities or configurations that constitute the potential inherent in that original wave function … a potential that is given realization, bifurcation by bifurcation, through the dynamics of various physical interactions or events.

Everett gave written expression to the foregoing ideas in the form of a thesis. John Archibald Wheeler was Everett's academic advisor.

Wheeler took Everett's written work to Copenhagen. He discussed Everett's ideas with, among others, Niels Bohr, the primary architect of the Copenhagen theory of quantum interpretation.

Wheeler tried to portray Everett's position as being consonant with the perspective of the Copenhagen theory of quantum interpretation. This meant, among other things, that Wheeler described Everett as being on-board with the Copenhagen approach to the issue of measurement and that Everett was merely trying to provide a more generalized account of the Copenhagen position.

However, a year, or so, later, Everett clearly indicated to, among others, Bryce DeWitt, the editor of *Reviews of Modern Physics*, that Everett believed the Copenhagen theory of quantum interpretation was not only incomplete but was also problematic in the manner in which that theory extended a substantial reality to the classical realm but withheld the same quality of substantiality from the quantum realm.

Because a variety of issues concerning his thesis remained unresolved, Everett took a research position with the Pentagon and moved from Princeton to Washington. However, from a distance, Wheeler continued to engage Everett and eventually induced Everett to cut the latter's thesis down to one-fourth of its original length.

Among the ideas that were eliminated from Everett's thesis during the process of downsizing were those that involved, among other things, the issue of ontological paths that could split off from one another and become part of a different realm of the universe. The eliminated ideas gave expression to issues that were most in conflict with the Copenhagen theory of quantum interpretation.

Everett's thesis committee accepted the revised, pared-down version of Everett's work. A few months later, in mid-1957, the truncated thesis was published in *Reviews of Modern Physics*.

In 1956, Wheeler had shown the full version of Everett's thesis to a physicist. Alexander Stern, who was at Bohr's Institute for Theoretical Physics. Stern referred to Everett's position as "theology", but this was sort of like a pot calling a kettle "black".

After all, the Copenhagen theory of quantum interpretation was very theological-like both with respect to its content (long on litanies, but short on evidence), as well as in relation to the evangelical-like fervor with which that theory was often promoted among physicists. For example, as mentioned earlier, Heisenberg had been pushed to tears by the unrelenting intensity through which Bohr pressed Heisenberg to reject Heisenberg's own position concerning the issues of uncertainty and measurement, and, as well, one might recall the way in which Bohr wouldn't leave a sick Schrödinger alone until the latter had, in some way, acquiesced to Bohr's perspective on various matters of interpretation.

Even Wheeler seemed to be cowed by, and appeared to feel the need to genuflect before, the Copenhagen style of catechism, As a result, he kept trying to re-package Everett's original ideas in a manner that might be viewed as being more acceptable to the scientific bishops and cardinals of Copenhagen as well as to their acolytes elsewhere.

Wheeler's foregoing actions might be understood in several ways. On the one hand, his efforts could be construed as being consistent with what a savvy, academic advisor might undertake to do in an attempt to try to assist his charge (i.e., Everett) to navigate through the dangerous waters of heterodoxy ... that is, Wheeler's actions could be seen as those of someone who knew the realities of the world of physics which prevailed at that time and, as a result, understood what might be accepted and what might prove to be a stumbling block as far as obtaining a degree is concerned.

On the other hand, Wheeler's actions could also be seen as those of someone who was serving as a defender of the faith. As such, his task would be to induce Everett, little by little, to conform to the accepted scientific theology of the day by getting him to jettison various ideas – such as splitting realities – that conflicted with the sort of Copenhagen orthodoxy that ruled the minds and hearts of many a physicist at that time.

In the end, Wheeler helped his advisee to get a degree. However, which of the foregoing scenarios best gives expression to the motivations underlying Wheeler's course of action is hard to know ... and, perhaps, both of the foregoing possibilities helped shaped how Wheeler interacted with Everett.

The fact that Wheeler tried to get Everett's original, longer work published by the Royal Danish Academy of Sciences suggests that Wheeler was intrigued by Everett's ideas. The fact that Wheeler traveled to Denmark in order to discuss those ideas with, among others, Bohr suggests that, on some level, Wheeler understood there were ideas present in Everett's work that might prove to be contentious as far as the ruling Copenhagen theory of quantum interpretation was concerned.

Moreover, the fact that Wheeler initially sought to obtain Bohr's blessing concerning Everett's work indicates that Wheeler knew the conceptual politics that prevailed in the world of physics at that time. In addition, the fact that Wheeler induced Everett to get rid of all – or most -- of the controversial aspects of the latter's ideas indicates Wheeler's acknowledgement that Bohr's philosophical influence was too pervasive and entrenched for Everett's ideas to be received with any degree of equanimity by the world of physics in the mid-1950s ... and, actually, for many years to come.

One might note in passing that Everett's experiences with respect to having to run an ideological gauntlet in order for his ideas to become acceptable to academia is not an anomalous set of circumstances. For the last thirty years, or so, those who have been running the physics departments in many universities in North America have forced an array of students, post-doctoral fellows, research participants, and the authors of various journal articles concerning quantum physics to bow down before the altar of string theory if such individuals wanted to get their degree, or someone wanted to get an article published, or if a researcher wanted to get grant money, and so on.

One of the values that often are promulgated in relation to science involves the idea that science is the quintessential form of searching for the truth in an impartial, rigorously objective manner. Unfortunately, there is considerable evidence (and the experiences of Everett and what has been happening with respect to the politics of string theory over the last 30 years, or so, constitutes just part of that evidence) indicating that such values are, sometimes, more of a myth than a reality when it comes to the teaching and practice of science.

Notwithstanding the foregoing considerations, Everett's many-worlds perspective appeared to give expression to a greater degree of consistency than did the Copenhagen theory of quantum interpretation. For example, unlike the latter perspective, Everett considered both the classical and quantum realms to be equally real or substantial, and, as well, Everett maintained that both the macro and micro levels of reality were governed by the same principle of superposition.

Nonetheless, whatever Everett's position might bring in the way of greater consistency did not mean that his perspective was devoid of problems of its own. Foremost among such problematic issues was the ontological character of bifurcation.

More specifically, the dynamics of the splitting process appear to be relatively vague if not mystifying. While Everett clearly indicates that each splitting event gives expression to different dimensions of the potential that is encompassed by the initial wave function, what is less clear is how each bifurcation or branching process generates a complex reality that provides an existential context in which a given dimension of the wave function's potential becomes manifest.

Suppose, for instance, there is a dimension of the wave function that -- when measured or when a given interaction occurs – is able to have two values – 'a' and 'b'. A measurement takes place or an interaction occurs, and, as a result, the wave function splits, and on one occasion, 'a' is expressed but not 'b', and on another occasion, the other value 'b' is expressed but not 'a'.

According to Everett, 'a' is ensconced in a separate, accompanying reality, and, as well, 'b' is also embedded in a separate, accompanying reality. How did these accompanying realities become part of the splitting process, and how did they acquire the properties that characterize them?

Apparently the splitting process creates two entirely intact and complete branches of reality. Each of those branches is the same as the other one except that in one bifurcation, 'a' occurs, while in the other bifurcation 'b' occurs,

How does the bifurcation process create complete worlds to accompany 'a' and, then, 'b'? Furthermore, why isn't it the case that possibility 'a' occurs at a given time in this world and, then, possibility 'b' occurs at a later time in this <u>same</u> world ... that is, why must separate worlds be generated?

What is the relationship between the splitting process and the possible configurations that are encompassed by the potential of the original wave function? Does that wave function get reproduced for each bifurcated world (and, if so, how), or does each bifurcated world have continuous access, in some unspecified manner, to the configuration possibilities inherent in the original wave function?

What is the nature of the process that keeps split worlds separate from one another? Where are these bifurcated worlds located?

What is the source of energy that enables an indefinite number of split worlds to be generated and maintained? How is that energy organized to create the world that accompanies the bifurcation process which gives expression to one dimension of the set of possibilities inherent in the potential of the original wave function rather than some other dimension of that same set of possibilities?

In addition to the foregoing issues, Everett's position shares a problem with the Copenhagen school of quantum interpretation. More specifically, neither of those perspectives can explain in what sense the principle of superposition constitutes an actual reality in the scheme of things.

From what dimensional realm does the principle of superposition go about governing the unfolding dynamics that take place in the universe? While one can understand the nature of configuration space and how the latter gives expression to all of the possibilities that are inherent in the potential of a wave function, one has considerable difficulty understanding why one should consider such possibilities to have a reality that is more ontologically substantial than just being a conceptual description of such possibilities, or why, even if such possibilities possess, in some sense, an ontological reality that transcends the purely conceptual, one must suppose that all such realities exist simultaneously.

The ontological reality of the principle of superposition needs to be demonstrated. Both Everett and the Copenhagen theory of quantum interpretation take the principle of superposition as an ontological given, and, consequently, neither of the foregoing perspectives puts forth any evidence indicating that such a "given" is ontologically warranted rather than merely being a heuristically valuable assumption.

Furthermore, even if one were to assume that the principle of superposition had some sort of ontological reality, why should one further suppose that the nature of the universe is exhausted by the particle dynamics that give functional expression to the possibilities that characterize configuration space? Is it necessarily the case that such phenomena as: Consciousness, reason, creativity, and understanding are strict functions of physics, and if this is not the case, then, the wave function that supposedly describes the physical nature of the universe is woefully incomplete.

The many worlds perspective often is characterized in a way that suggests that everything that could happen will happen. However, just because a possibility exists does not necessarily guarantee that such a possibility must become realized.

The probability distribution that represents the likelihood that a certain quantum possibility will occur is not a reflection of the character of past events. The probability distributions associated with a given wave function, along with Schrödinger's equation, are not part of a statistical technique for organizing data in the form of some sort of frequency table, but, rather, that distribution constitutes a prediction about what is likely to be observed under a given set of conditions.

There is no necessity associated with the foregoing sort of probability distribution that ensures or requires that each and every one of those probabilities must occur sooner or later. Quantum probability distributions indicate the array of possibilities that might occur on any given occasion but say nothing about whether all such possibilities must, at some point or another, be realized.

Consequently, there is nothing in the wave function, or the principle of superposition, or Schrödinger's equation that is capable of justifying a claim often associated with Everett's many worlds theory that everything that can happen will happen. Every bifurcation or splitting process that occurs is an expression of the dynamics inherent in a given set of conditions, and there is nothing in those dynamics that necessitates that every dimension of the associated wave function will -- sooner or later -- become manifest.

In short, possible probabilities for a given set of conditions can be calculated. However, while ontological reality might gives expression to some of those possibilities, nothing in Everett's perspective demonstrates that all such possibilities must necessarily become manifest.

The picture entailed by Everett's many worlds approach to understanding quantum phenomena is as elusive and illusory as is the portrait painted through the Copenhagen theory of quantum interpretation. Despite the mathematical wherewithal and scientific trappings that can be called upon to frame and inform those two perspectives, nevertheless, at the present time they are both just exercises in philosophical hermeneutics rather than actual science.

In time, the foregoing perspectives might be demonstrated to be something more than an exercise in philosophical hermeneutics. However, the significance of the foregoing use of "might" is to state a possibility rather than to allude to any sort of necessarily realizable probability.

What is the nature of reality? Although physicists have accomplished a great deal as far as being able to mathematically (and, very frequently, quite accurately) describe the behavior of physical systems and how various properties associated with such systems change over time, physicists are not necessarily able to give any straightforward answer to the foregoing question concerning the nature of reality.

Physicists use the Standard Model of physics on a regular basis. The Standard Model enables physicists to predict, in extremely precise terms, how a number of elementary forces will interact with different kinds of particles under a variety of conditions.

In certain ways, however, the foregoing manner of talking about the dynamics of quantum entities is somewhat illusory. For example, one of the methodological backbones lending conceptual support to the Standard Model is quantum field theory.

Every 'particle' is described through the properties of the quantum field that is associated with any given particle. Among such properties is one that indicates that unlike a classical field, a quantum field is discrete rather than continuous, and, in addition, one cannot necessarily identify a particular location in that field that gives expression to a discrete particle with a definite momentum.

If a field is discrete, how are different influences propagated through that field? Ontologically – and not mathematically -- how are different loci in such a field related or linked with one another?

| Quantum Queries |

Do particles generate a quantum field? Or, does a quantum field give expression, under certain conditions, to phenomena that exhibit a set of properties that, under different conditions, are referred to as being particle-like and/or wave-like?

Is a vacuum – which by definition is, supposedly, devoid of particles – as empty as its definition would suggest? Or, could it be that while a vacuum might be empty of particles, nevertheless, a vacuum gives expression to a field that, under the right circumstances, is capable of giving expression to particle-like events?

Operators – which are mathematical functions – provide a way to describe the behavior of quantum fields. However, what actually might be responsible for generating and shaping such behavior tends to fall beyond the purview of those operators.

Operators don't operate on a physical field, per se. Operators interact with another mathematical entity known as a state vector that gives expression to various possible quantum configurations for a given set of conditions.

Together, mathematical operators and state vectors generate probabilities. Those probabilities have more to do with predicting the likelihood that certain kinds of behavior might occur under a given set of conditions rather than constituting an explanation of what the nature of the reality is that underwrites such probabilities.

A quantum field is the mathematical representation of the dynamics that are possible in a given context. While the behavior of the dynamics being described might conform to the mathematics of quantum field theory, whether, or not, that which is being described in the foregoing way is, ontologically speaking, a field, and/or a particle, and/or a wave is not necessarily clear-cut.

Being field-like, or particle-like, or wave-like is not necessarily the same thing as being a field, or a particle, or a wave. Reality might appear to be many things while actually being something else altogether ... a 'something' that makes appearances with such particle-like and wave-like and field-like properties possible.

Some physicists believe that sets of properties and relations are all that exist. In other words, according to such physicists, when one looks at the world, one doesn't see objects per se ... instead one sees a set of relationships or a set of properties that one learns to label as being this kind, or that kind, of object.

Various terms are used in conjunction with the foregoing perspective. One such term is "trope ontology," and another term is "ontic structural realism".

Whatever the term is that might be used to give expression to the foregoing perspective, the general approach embodied by such terms doesn't seem to have anything to offer which is all that different from the traditional terminology of "field", " particle", and "wave". One can ask the same question of 'relations' and 'sets of properties' that one can ask of 'fields', 'particles', and 'waves' ... that is, what makes them possible?

Relations and sets of properties aren't necessarily any better at accounting for the nature of their own existence – or how such an existence is possible -- than are fields, particles and waves. There seems to be an illusory, as well as an elusive, dimension to all of these terms that doesn't necessarily permit one to gain a clear understanding of the reality whose behavior is being so precisely described, at least to some degree, by the Standard Model of physics and that model's associated methods of mathematical representation.

Some scientists are not bothered by the foregoing considerations because they don't believe that the purpose of science is to determine the nature of reality. Instead, they feel that the essential nature of science is to serve as a vehicle for generating predictions and hypotheses that can be tested ... although I am not quite sure what the value of a prediction or a hypothesis is if, after being vetted through a testing process, such a prediction or hypothesis doesn't have something to do with disclosing or uncovering an aspect of the nature of reality.

In any event, viewed in the foregoing way, science becomes a conceptual means of transporting a scientist from one hypothesis to the next in an empirically rigorous and demonstrable fashion. As such, science appears to be more of a moment-to-moment struggle about trying to differentiate between defensible hypotheses and problematic ones rather than trying to discover the deeper nature of reality.

Of course, not all scientists think about science in the foregoing fashion. Many scientists feel that science does have something to say about the nature of reality ... although what science has to say in this regard tends to be the result of a complex, time-consuming process that reaches limited and tentative, but heuristically valuable, conclusions concerning reality's possible nature.

However, even if the earlier characterization of science (the one which says that science is a means of generating predictions and hypotheses that can be tested) were correct – and I am not sure how one would go about differentiating between such a perspective and a point of view that treated science as being nothing more than an arbitrary expression of personal likes and dislikes as far as fixing the meaning of science is concerned – nonetheless, the earlier notion of science does not seem to serve the interests of anyone who might be seeking to establish a defensible or plausible way of engaging the reality problem ... at least not directly -- although, perhaps, in light of the earlier characterization of science (two paragraphs back), such an activity still might be able to indirectly serve the interests of those who were interested in discovering the nature of reality if one were to critically inquire into the implications, if any, that such activity might carry for a project that sought to travel beyond instrumental horizons that were limited to testing a given hypothesis or prediction.

| Quantum Queries |

Chapter 7: Physical Conundrums

Quantum electrodynamics (QED) gives expression to a quantum field theory of electrodynamics that takes into account the principles of special relativity ... an idea that began with the work of Paul Dirac in 1927 and, eventually, was brought to the first stage of completion through the work of Hans Bethe, Richard Feynman, Julian Schwinger, Freeman Dyson, and Sin-Itiro Tomonaga. In general terms, quantum electrodynamics describes the way that light interacts with matter.

One of the methodological means through which scientists test and probe quantum electrodynamics is by analyzing the dynamics that occur in simple atoms such as hydrogen. Physicists can explore various properties of QED by examining certain kinds of energy shifts that occur in hydrogen atoms.

For example, Willis Lamb, Jr. initially discovered one form of the foregoing kind of energy shift in 1947 while conducting experiments involving hydrogen. A "Lamb shift" is shaped by several factors.

One of those shaping factors concerns the dynamics of virtual particles. Virtual particles supposedly appear and disappear within – among other places – atoms, and QED provides a means of calculating, with considerable precision, the extent to which the dynamic of virtual particles affects the energy shifts that were first observed by Lamb.

The foregoing dynamic is described in terms of exchanges of energy that, supposedly, are introduced through virtual particles. However, virtual particles might only constitute a terminological way of referring to the aforementioned dynamic, and, virtual particles, per se, might, or might not, exist.

To be sure, 'something' is affecting the way the Lamb energy shift is manifested. Moreover, whatever that 'something' is, QED is capable of capturing its behavioral properties with great precision.

Nonetheless, the foregoing 'something' does not necessarily need to be a function of virtual particles. Rather, the idea of virtual particles could be just a stand-in term that is used to acknowledge the presence of a capacity or force – in, for example, a hydrogen atom – that is able to impact the nature of a Lamb shift.

A second factor that can shape the character of a Lamb shift is a function of both the size of a proton's radius as well as the manner in which an electron occupies an orbit. According to quantum mechanics, an electron does not occupy a particular location within an atom but, rather, one needs to square the wave function for the electron in order to establish a probability distribution for finding an electron in one quantum state at a given location within the atom, as opposed to finding that electron in a given quantum state at some other location within that same atom.

Quantum mechanics maintains that one of the possibilities for an electron's location is within the proton itself. If an electron were inside of a proton, the strength of its interaction with the proton would not be as great as that coupling strength would be if the electron were outside of the proton ... It would be less by a factor of 0.02 percent.

The size of the proton's radius will affect the foregoing considerations. The greater the size of the proton's radius, the greater will be the probability that an electron might be found within a proton, and, therefore, the more likely that the coupling strength between a proton and an electron might be diminished under such circumstances.

Until about five years ago, physicists believed they had a fairly good understanding of the proton, including its size. The foregoing belief has been called into question by a series of experiments.

Prior to 2010, the radius of a proton had been deduced from data involving the spectroscopic measurement of energy levels in hydrogen atoms, along with other experiments that generated information concerning scattering properties of electrons that had been fired into hydrogen gas. In the latter case, the greater the deflection that is observed, then, the shorter the wavelength will be that is manifested and the more energy will be involved, whereas, the smaller the deflection that is observed, then, the longer the wavelength will be that is generated, and the energy involved will be less than in the former case. Information concerning the radius of a proton is found by analyzing the longest of the foregoing wavelengths.

All of the foregoing experimental results gave – within allowable degrees of experimental error -- consistent answers with respect to the size of a proton. The radius of the proton was computed to be approximately 0.877 femtometers (a femtometer is 10^{-15} meters).

Beginning in 1997, however, a series of experiments (led by Randolf Pohl) began to be set in motion that involved an alternative approach to measuring the Lamb shift and its relationship to the size of a proton's radius. These experiments involved muonic hydrogen ... that is, hydrogen atoms in which the electron is replaced by its more massive relative (by a factor of 200): The muon.

The greater mass of a muon permits it to get roughly 200 times closer to a proton than an electron can achieve. At the same time, the muon was calculated to have a significantly increased likelihood (by a factor of eight million) of spending more time within a proton than is the case for an electron.

This greater probability of a muon spending time inside of a proton would impact the Lamb shift. As indicated previously, the size of that impact is approximately 2 percent.

The general form of the experiment was as follows. Muons from an accelerator were to be beamed into a container filled with hydrogen gas.

| Quantum Queries |

From time to time, a muon from the accelerator beam would replace an electron in a hydrogen atom. When this occurred, an atom of muonic hydrogen would form that existed in a highly excited state (known as 2S) and, then subsequently (within several nanoseconds), would fall into lower energy states.

When a muon entered the hydrogen-containing vessel, a laser kicked into action. If the energy (wave length) of the laser were exactly right, it would push the energy state of the hydrogen atom up to 2P, and this would be followed by a fall in energy down to the 1S state.

If the laser beam did not contain precisely the amount of energy that marked the difference between the 2S and 2P energy states, nothing seemed to happen. However, if the laser gave expression (in the form of a wave length of the right size) to exactly that energy difference, then, a low-energy X-ray photon could be observed (a function of the fall in energy from the 2P state to 1S), and, as a result, the experimenters would know which energy state was present.

In a hydrogen atom that is in a 2p energy state, a muon is calculated to spend zero time inside of the proton of the hydrogen atom. Since 2S and 2P states involved energy differences, by analyzing those differences, one could make inferences about how much time a muon spent within a hydrogen atom's proton and, therefore, how much the Lamb shift energies were likely to be affected.

Although the aforementioned hydrogen experiment has been described – at least in general terms – in just a few paragraphs, the time required to actually get such an experiment going, properly calibrated, and so on, can take years.

The foregoing idea was put forward as a proposal in 1997 in the hope of being granted time on the accelerator at the Paul Scherrer Institute in Switzerland. The proposal was accepted nearly two years later.

| Quantum Queries |

A further three years were required to: Construct the right kind of laser system; build the detectors needed to register the presence of the low-energy X-ray photons that would be generated when the lasers were able to push muonic hydrogen from a 2S energy state to a 2P energy state, and, finally, generate a beam of low-energy muons. Another year was eaten up getting the experiment ready for the actual run in 2003, and this involved having to overcome a number of technical problems.

Following three weeks of data generation in 2003, the researchers didn't find the signal for which they were looking. The proton radius didn't seem to be anywhere in sight.

One of the first conclusions drawn in relation to the foregoing results was that there must be something wrong with the experimental set-up. Consequently, the experimenters decided to revamp the laser system, and, this required an additional three years of work.

Three more weeks of data were collected in 2007. Once more, the experimenters did not observe the signal for which they had been looking.

Another run of data was generated in 2009. Again, no signal was observed.

Several final runs were conducted in 2009. However, on these occasions, the decision was made to look for the sought-for signal in a more restricted space – as if the radius of the proton were smaller than it previously had been calculated to be.

On July 4, 2009, a clear signal was detected. When the radius of the proton was measured in conjunction with muonic hydrogen, apparently the radius was smaller than anticipated. The measurement indicated that the radius of the proton was 0.8409 femtometers, some four percent smaller than previous measurements (approximately 0.0877) had indicated.

The foregoing experiment was ten times more accurate than any previous experiment had been that sought to measure the radius of the proton. Yet, this – apparently – more accurate measurement indicated that the radius of the proton was smaller than previously believed.

Performing the calculations that are necessary to analyze laser wavelength measurements and reconfigure the latter in the form of a proton radius size is a fairly complex process. A number of physicists wondered if some sort of error had taken place in relation to that translation process.

Consequently, the calculations were done again -- and in expanded ways -- by a variety of scientists. The original calculations were verified.

Four years have passed since the Pohl experiments were completed. The difference between the results of those experiments and the previously accepted value for the radius of a proton remain.

Another problem – possibly related to the foregoing issues – has arisen in conjunction with the muon. The muon – like its electron cousin – has a magnetic moment (the torque force that will be experienced by a muon in a magnetic field). The measured value of that magnetic moment does not reflect the value of the magnetic moment that is calculated by means of QED theory.

Different theories (e.g., the presence of previously undetected particles that would induce muons to behave differently than electrons do) have been proposed to account for the foregoing sorts of conundrums. However, to date, physicists have had difficulty envisioning any kind of theoretical particle that would explain the foregoing results without simultaneously having observable effects in relation to other experimental results ... effects that, so far, have not been observed.

Perhaps, the phenomena that are referred to through the terminology of 'virtual particles', along with the energies that give expression to the Lamb shift, as well as the magnetic moment of the muon, and, finally, the radius of the proton have some sort of underlying set of dynamic connections. Various kinds of experiments are being discussed to see what, if anything, might be discovered that is capable of illuminating those issues.

For instance, to date, scattering experiments have only been done with electrons, and, therefore, some physicists want to see what would happen in muon scattering experiments and measure those against the results from electron scattering experiments. In addition, some physicists wonder if the same kind of smaller proton size will turn up if one runs similar experiments using deuterium (consisting of a proton and a neutron).

From initial conception to final data analysis, the Pohl experiments took 12 years to complete. Presently, those efforts are suspended somewhere between, on the one hand, the possibility that the tip of an iceberg involving new physics might be making its presence known through such results and, on the other hand, the possibility that subsequent experiments will discover that some currently unknown set of errors – experimental or otherwise – will undermine the Pohl findings concerning the size of the proton radius.

I admire the vision, ingeniousness, technical wherewithal, calculating wizardry, persistence, and patience of experimental physicists. Nevertheless, there are a variety of problematic issues that also occupy the mental context through which such admiration arises.

Twelve years of continuous effort is a long time to spend to end up in the midst of uncertainty concerning the nature of reality ... especially given how small that portion of reality is. If one extends the foregoing temporal framework to encompass all of the original experimental efforts that went into establishing a value for the radius of the proton (a value that had been accepted for many years prior to Pohl's muonic hydrogen experiments), then nearly a quarter of a century, or more, has been consumed in trying to determine one thing: The size of a proton's radius ... and at this point, the results are mixed – both for scientists and the lay public.

The foregoing risks, problems, frustrations, failures, and rewards are all part of the process of doing experimental science. However, those considerations don't necessarily advance the needs very far of those people who are not scientists and, consequently, the latter individuals continue to be faced with the choice of how to best spend their time with respect to engaging the reality problem.

Without wishing to diminish any of the remarkable work that has been done and is being done by physicists, and notwithstanding that scientists will, very likely – and sooner, perhaps, rather than later – determine whether Pohl was right or wrong, or determine whether, or not, virtual particles, the Lamb shift, the radius of a proton, and the magnetic moment of muons all have something to do with one another, nevertheless, it seems rather obvious that non-scientists aren't necessarily in a very good position to spend their lives pursuing issues that will not necessarily get them much closer – except in very limited ways – to figuring out how to engage the reality problem in the time that is available to them.

The size of a proton's radius is a conundrum for scientists. Life is the primary conundrum for human beings, and solving the proton problem will not necessarily contribute a great deal toward solving the conundrum of life.

In physics, the 'generation problem' refers to the existence of three levels of particles among quarks and leptons, and, yet, only one of those generations of particles tends to dominate the world beyond the horizons of accelerators and colliders. For example, in the natural world that exists outside the artificial manipulations of human experimental intervention, protons and neutrons seem to be constructed from just two kinds of quarks – namely, 'up' and 'down' quarks, whereas, for the most part, leptons in the everyday world of nature appear to consist of just two particles – electrons and electron neutrinos.

There are exceptions to the foregoing 'normal' course of events that, currently, are not fully understood. For instance, under certain circumstances, neutrinos appear to have the capacity to switch identities among the three generations of neutrinos ... thus, all three generations of neutrino leptons (i.e., electron, muon and tau neutrinos) participate – according to some still-to-be specified set of principles – in the foregoing transformation process.

The second- and third-generation quarks (charm and top) that share the same charge as the first-generation up quark (i.e., +2/3) become heavier with each generation. However, aside from the differences in mass, the 'up', 'charm' and 'top' quarks appear to be pretty much the same as one another.

Similarly, the second- and third-generation quarks (strange and bottom) that share the same charge as the first-generation down quark (i.e., -1/3) also become more massive with each succeeding generation. Nonetheless, once again, in all other respects the three generations of quarks that share the same -1/3 charge appear to be the same.

The heavier editions of the 'up' and 'down' quarks are all unstable. Within a very short time after becoming manifest, the heavier, 2nd- and 3rd-generation quarks all decay into their first-generation counterparts.

Second- and third-generation leptons – namely, the muon and tau particles also are unstable ... dissipating, respectively, in about 2.2×10^{-6} and 2.9×10^{-13} seconds. However, unlike 2nd- and 3rd- generation quarks, such later-generation leptons do not decay into just a first-generation particle (i.e., electron or neutrino) but, instead, give expression to an array of possibilities (more complicated in the case of the heavier tau particle than is the case with respect to the muon particle ... indeed, the heavier tau particle is the only lepton that can, under certain conditions, decay into a hadron – that is, particles that are sensitive to the presence of the strong force).

Although muons can be generated through the high-energy physics of cosmic rays, nonetheless, for the most part, muons don't appear to have any function in normal matter. Consequently, physicist, I.I. Rabi, is reported to have uttered: "Who ordered that?" when he found out about the discovery of the muon.

With the possible exceptions associated with the identity-switching phenomenon involving second- and third-generation neutrinos that was noted earlier (I.e., muon and tau neutrinos), Rabi's aforementioned sentiments could be uttered with respect to most of the second- and third- generation particles. Such later-generation particles don't seem to play much of a role, if any, in what might be referred to as "normal physics", and why this should be the case is something of a conundrum for physicists.

One approach that tries to account for the existence of differences among the foregoing generations involves the notion of "preons". The latter term is a generic way of referring to various possibilities that involve hypothetical, constituent components of quarks and leptons.

The Standard Model of physics treats quarks and leptons as being point particles that are devoid of any sort of internal structure. However, what appears to be a unitary point at one level of scale might give expression to a more complex internal, structural arrangement when observed on a smaller level of scale.

Thus, being able to observe the possible internal structure of quarks and leptons might be a matter of achieving better resolution. Just because current capabilities (in the form of colliders and accelerators) are not able to resolve the internal structure of a quark or lepton (if such an internal structure exists), those sorts of limitations don't mean that more powerful accelerators and colliders might not be able to bring those possibilities to a level of resolution that can be observed either directly or indirectly.

Preons are hypothetical, fermion entities (matter) that refer to the possible constituents of particles – namely, quarks and leptons – that are considered to be just unitary, point particles at the present time. Such hypothetical particles have been given different theoretical properties by various theorists.

For example, in 1979, Michael Shupe and Haim Harari proposed that there were two kinds of preons, each with an antiparticle. One preon had an electrical charge of -1/3 (and, therefore, its antiparticle had an electrical charge of +1/3), while the other preon had an electrical charge of 0.

Quarks and leptons consisted of different three-preon combinations. Thus, an up quark consists of two, +1/3-charged preons, together with one preon of 0-charge, whereas the antiparticle for such an up quark is made up of two, -1/3-charged preons, along with a 0-charged preon.

Unique, three-member preon configurations were worked out for quarks, leptons, and bosons. Harari and Shupe (along with Harari's student Nathan Seiberg) used such combinatorics to describe the sorts of interactions that currently are described by the Standard Model in relation to first-generation particles.

Extending the foregoing preon model to 2nd- and 3rd- generation quarks and leptons becomes a much more complicated and problematic affair. While there are many details in their model that have not, yet, been worked out, the bottom line is for Shupe and Harari is that the more massive generations of quarks and leptons were considered to be energized versions of first-generation quarks and leptons.

Even if the aforementioned theorists could work out a consistent system of preon combinatorics for: All three generations of particles (together with their antiparticles), as well as the known bosons (force carrying particles ... photons for the electromagnetic force, gluons for the strong force, W and Z particles for the weak force, and the Higgs particle or particles for the property of mass), nonetheless, such a model might resolve the generation problem only at the cost of creating other kinds of problems. The latter would include all the difficulties (many of which have not, yet, been worked out even theoretically -- let alone experimentally) that are entailed by the extra level of preon particles that are being hypothesized to account for why three generation of particles exist.

For example, just as gluons hold quarks together within neutrons and protons, one might suppose there must be some kind of boson (or force-carrying particle) that holds preons together within the quarks and leptons for which the preons are constituent parts. However, the identity of such a possible boson – if it exists – stands in need of discovery.

Notwithstanding the foregoing sorts of problems, other questions remain. For instance, while claiming that second- and third-generation particles are merely more energized versions of their first-generation counterparts does offer a plausible way to account for why there is no actual generation problem since such 2nd- and 3rd-generational states are only likely to occur under the extreme conditions of accelerators/colliders (or, possibly, cosmic ray dynamics), nevertheless, does one need to resort to an added, epicycle-like notion of preon sub-particles in order to be able to suggest that 2nd- and 3rd- generation particles are merely more energized versions of the first-generation counterparts?

Conceivably, 2nd- and 3rd-generation particles are more energized editions of first generation particles due to the presence of some sort of unknown force (and sans preon particles). For example, perhaps there is more than one kind of Higgs field (or some other kind of field of force) that interacts with particles, and depending on the nature of such a field, one might get more-massive or more energized versions of the first-generation particles, and, in addition, such fields might only manifest themselves under certain conditions of high-energy physics.

As such, rather than being an inherent -- but inexplicable -- part of the structure of the universe, 2nd and 3rd generation particles would be an artifact of the high-energy physics of accelerators and colliders ... and, on occasion, would be given expression through the high-energy dynamics of so-called cosmic rays. Consequently, one doesn't necessarily have to resort to the added complexities of a preon model in order to argue that 2nd- and 3rd-generation particles are merely more energized editions of first-generation particles.

Of course, there is a preon-based response to the foregoing possibility. More specifically, different configurations of preons might be the reason why the Higgs field differentially affects the mass of various particles.

In other words, from the perspective of a preon model, there could be some dimension, property, or quality associated with certain preon configurations that constitute 2nd- and 3rd- generation particles that induce the latter to be more sensitive to the presence of a Higgs field and, as a result, renders those higher generation particles more massive than their first-generation counterparts. Assuming that preons constitute an accurate picture of the inner life of quarks and leptons, then, the problem becomes one of trying to figure out what the nature of the foregoing sort of difference-making dimension, property, or quality might be ... both theoretically and experimentally.

One of the problems associated with preon theories, is that they tend to give expression to what are referred to as confinement models. In other words, just as quarks are confined within, say, protons or nucleons, so too, preons are confined within quarks and leptons.

A feature of confinement models is that the masses of the particles being confined are inversely proportional to the size of the space within which confinement occurs. Given that leptons and quarks are much smaller than, for example, protons or neutrons, then if one extends the foregoing inverse proportionality relationship to preons, then, one is left with the possible conclusion that preons might be more massive than the protons or neutrons in which they are located.

Some theories have been worked out (for example, by Gerard 't Hooft) that might provide a way to overcome the foregoing problem. However, such theoretical possibilities have not, yet, been confirmed experimentally.

Preons, assuming they exist, might not constitute the ultimate foundations for fermions such as leptons and quarks. Some theorists believe that preons might be made from superstrings

If quarks and leptons do have an internal structure, then, the preon and/or superstring models might be viable candidates to account for the generation problem. On the other hand, if quarks and leptons do not have an internal structure, then, one must come up with some other explanation for why three generations of particles exist when, for the most part, only one of those generations tends to manifest itself in the natural world.

To date, all experimental results indicate that, for example, quarks are point particles that either have no size or have a size that is not greater than 10^{-18} meters. All one can do is experimentally push the horizons downward as one searchers for a possible nonzero value with respect to the size of a quark, and such a search needs to continue until one can generate data indicating that either quarks are point particles with no internal structure (and, therefore, are not a function of preon or superstring dynamics), or that quarks have a nonzero size which allows for the possibility that such particles are composite in nature and, therefore, are made up of preons, superstrings, or some other kind of sub-particle structure.

If it turns out that quarks are point particles (that is, they have a zero size), there are other issues that need to be addressed. For example, one needs to address the issue of how – without size --

quarks have the properties they do (e.g., electrical charge, mass, spin, color charge, property of confinement, sensitivity to the presence of the strong force) … that is, one needs to explain how something that is without size has the capacity to give expression to the foregoing array of properties, and, as well, one needs to account for how six flavors of quarks – each of which is without size – gives expression to differences in such properties

When the reconstituted LHC at CERN in Switzerland begins operations again in 2015, the projected energies of collisions will be between 13-14 TeV (trillion electron volts). Whether such energy levels will be able to generate data that sheds light on any of the foregoing issues remains to be seen.

According to Einstein's General Theory of Relativity, space has the capacity to act as if it were a deformable surface, and gravity is a force whose presence is indicated through the manner in which space is deformed. The foregoing perspective assumes that space is something that can be deformed and assumes, as well, that the nature of space is receptive to, or sensitive to, the presence of gravity.

Space might not be the deformable entity that Einstein supposes it to be. Whatever deformations take place in conjunction with the presence of gravity might have more to do with the way in which the gravitational field interacts with itself and other possible fields [e.g., electromagnetic fields, the Higgs field(s)] than such deformations have anything to do with the nature of space.

However, for the moment, let's assume that Einstein is right with respect to the foregoing issue. Let's assume that space is deformable and sensitive to the presence of gravity and proceed from that point – at least for now (This issue will be encountered again in Volume III of Final Jeopardy).

Both Newton and Einstein argued that all objects in the universe attract one another with a force that is proportional to the product of their masses and inversely proportional to the square of the distance between those objects. Consequently, the material contents of the universe should be inclined, to varying degrees, toward being mutually attracted to one another.

At the time that Einstein developed his General Theory of Relativity, scientists believed the universe to be a static entity. However, if the basic tendency of all bits of matter is to be attracted toward other bits of matter, then, how could the universe remain static?

To account for how a static universe could be reconciled with the presence of a universally attractive force that pulled material things toward one another, Einstein introduced: λ, lambda, a force that was capable of counteracting the presence of gravity and that, subsequently, came to be known as the cosmological constant.

In general terms, the value of lambda was believed to be in equilibrium with the force of gravity. However, depending on how the force of lambda interacted with the force of gravity at any given point in space, the general condition of equilibrium could be destabilized locally.

A few years later (1929), the astronomer, Edwin Hubble, made a discovery that raised questions about the significance of Einstein's added cosmological constant. Apparently, the universe was expanding -- not static – and, therefore, there was no need to posit the existence of something that served as a countervailing force to the attractive force of gravity in order to keep the universe in a static condition ... although if engaged from a different perspective, one might be able to use lambda as a way of accounting for how the universe, as a whole, would -- depending on how the value of lambda and gravity played off against one another -- either expand, stay the same, or collapse.

The lambda force (if it exists) and gravity engage one another in a localized manner. Quantum field theory also operates in a localized fashion.

For instance, according to quantum field theory, the so-called vacuum of space is not devoid of particles. Instead, the aforementioned field theory claims that energy fluctuations are present in the vacuum of space, and by means of such fluctuations, virtual particles (of all descriptions) blink into and out of existence within about 1×10^{-21} seconds.

Virtual particles are not necessarily directly observable. Rather, their existence is inferred by the way in which observable phenomena -- for example, the energy levels of hydrogen atoms -- are affected by the postulated presence of such particles (The opening section of the current chapter touched on the foregoing issue).

Certainly, the energy levels of hydrogen atoms seem to be sensitive to the presence of some form of dynamic. The latter form of dynamic might be a function of the activity of virtual particles, or that dynamic could be a function of some other kind of phenomenon, but in either case, one is uncertain about whether, or not, such dynamics are transpiring at every point in space and whether, or not, those dynamics necessarily generate every kind of particle.

Quantum field theory is part of a system of mathematical description that can generate predictions concerning the likelihood that certain kinds of properties might manifest themselves at given locations. As has been pointed out previously on a number of occasions, that theory doesn't have anything to say about the underlying nature of reality ... instead, it describes behavioral possibilities for a physical system under certain conditions.

The amount of fluctuating energy that is contained at any given point in the vacuum of space is unknown. Moreover, the extent to which, and the way in which, the vacuum of space fluctuates is unknown.

Irrespective of what the specific nature of the foregoing unknowns might be, Einstein's famous equation – $E=mc^2$ –

indicates there is an equivalency relationship between mass and energy. Therefore, to whatever extent energy is present in the vacuum of space, that presence also gives expression to an equivalent value in mass that has relevance to Einstein's General Theory of Relativity.

Quantum field theory predicts that every kind of virtual particle makes a contribution to the total energy of the vacuum. Furthermore, quantum field theory predicts there will be such an extensive array of high-energy particles present in the vacuum that the total energy will be infinite in nature.

The evidence of history has demonstrated that theories generating infinite results should be entertained with some degree of caution (e.g., remember the ultraviolet catastrophe – involving infinite results -- with respect to black body radiation that helped launch quantum mechanics). The presence of infinities generally means there is something wrong with the theory out of which such infinite values arise.

One way in which physicists have whistled their way past the vacuum energy cemetery is to ignore certain kinds of high-energy possibilities that have been predicted and, then, work with what remains. When this is done, physicists can disentangle themselves – at least in this case -- from the issue of infinity, but another problem bubbles to the surface.

More specifically, if one were to calculate -- just in relation to virtual photons -- the energy density of the vacuum in terms of joules per cubic centimeter, then, quantum field theory predicts that one will find something of the order of 1 times 10^{116th} joules of energy in every cubic center of the vacuum.

When one introduces energy density estimates for other kinds of virtual particles, the calculations become more complex. This is because various fermions and bosons can give expression to different kinds of energies (negative and positive) within the vacuum.

While some oppositely charged virtual particles might end up cancelling one another, what remains after everything that is <u>known</u> is taken into consideration is a mess. Even after arbitrarily jettisoning certain kinds of high-energy possibilities from consideration, the foregoing energy density estimates for the vacuum are still so high that if such a condition actually prevailed, the current universe could not exist ... and, yet, the universe does exist, and, therefore, there is something wrong with the way in which such estimates are made or in which various calculations are done in conjunction with that issue.

One theoretical way of trying to resolve the energy density issue associated with quantum field theory's characterization of the vacuum is in the guise of theories involving some sort of Supersymmetry. Among other things, such theories do not exhibit any vacuum energy because they provide a means through which bosons and fermions are able to cancel one another out as far as the vacuum is concerned ... although just because there are technical ways of accomplishing such cancellations that make sense mathematically, one cannot automatically assume there are real world counterparts to those mathematical techniques that are capable of accomplishing such cancellations that can be proven to exist or that make much sense ontologically speaking.

Supersymmetric models are completely hypothetical in nature. To date – and this might change in the light of data that could be produced through the LHC experiments to be run in 2015 (and beyond) at CERN – there is no evidence that bosons and fermions pair off in all the ways that would be required to be able to cancel one another and, thereby, eliminate the vacuum energy problem.

Another possibility – also hypothetical – involves the following idea. However successful quantum field theory might be with respect to making predictions concerning the behavior of particles under certain conditions, nonetheless, that theory constitutes a poor model for predicting what might, or might not, be transpiring in the vacuum.

In other words, perhaps, the reason why quantum field theory gives such ridiculous estimates for energy density values in the vacuum is because the vacuum doesn't operate in the way quantum field theory assumes that the vacuum does ... that is, in accordance with the principles of quantum field theory. For instance, even if virtual particles exist (and they might not exist ... although there does seem to be some sort of dynamic occurring in the vacuum at scales below what is currently resolvable through available methodologies), nevertheless, virtual particles are not necessarily generated in the vacuum to the extent, or in the way, or with the energies that are predicted by quantum field theory.

In short, quantum field theory claims that the vacuum of space gives expression to all manner of energy fluctuations. Maybe, the vacuum doesn't consist of fluctuating energies even though, depending on local circumstances, certain kinds of energies might fluctuate from time to time and place to place, but such fluctuations might not occur as a function of quantum field theory but, rather, they take place as a function of what the nature of the vacuum is and what it permits – and doesn't permit -- in the way of such fluctuations.

Quantum field theory appears to be imposing a manner of characterizing the vacuum that is not appropriate. One of the consequences of such a problematic characterization involves the calculation of energy densities that are totally out of whack with what can be observed in conjunction with the vacuum ... namely, that the sorts of energy densities that are predicted by quantum field theory in relation to the vacuum do not seem to exist in nature.

When a theory makes predictions that don't reflect the available data, then, one needs to question the viability of that theory. Under such circumstances, either the theory needs to be jettisoned altogether as a model for that to which it is referring (in this case, the vacuum), or that theory needs to be substantially revised in a way that will assist it to become reconciled with that portion of reality for which it is currently making such problematic predictions.

The issue of the vacuum's energy density is a physical conundrum for quantum field theory. Maybe that conundrum will disappear in the near future – for example, if some version of Supersymmetry turns out to be viable – but for now, the foregoing issue remains, as it has been for more than half a century, an unsolved problem.

Einstein's special theory of relativity rests on two assumptions. One of those assumptions is rooted in the work of Maxwell, while the other assumption hearkens back to Galileo.

Maxwell's equations predict that the speed of light is a constant. Einstein accepted this prediction at face value and was prepared to follow the implications of that prediction to its logical conclusions ... namely, that the speed of light is independent of the relative motion associated with either the source of such light or an observer of that light.

The second assumption adopted by Einstein had to do with Galileo's ideas about whether, or not, one could empirically establish the existence of absolute motion. Galileo believed no one would ever be able to conduct an experiment capable of detecting the presence of absolute motion, and Einstein was in accord with that perspective.

Given the foregoing two assumptions, Einstein proceeded to explore what might ensue from such a starting point. What he discovered was special relativity.

Suppose one is in possession of a very simple kind of light clock. The clock consists of two mirrors that are placed one meter apart and which reflect light back and forth between them.

Every time light reflects off one of the mirrors, a detector attached to each mirror registers the impact and updates the temporal display by one unit of time. Since the speed of light consumes 299,792,458 meters per second, the time it takes for light to travel from one mirror to the other (separated by a distance of one meter) takes about 6.67 nanoseconds (6.67×10^{-9} seconds).

If the foregoing light clock is placed in, or on, something that is moving – say a flatbed, railroad car – what does an observer standing a short distance from those tracks see when that clock passes by. Prior to Einstein, everyone (including physicists) believed that an observer would see the clock tick at the same rate irrespective of whether the clock was stationary or moving ... namely, one tick per 6.67×10^{-9} seconds.

However, in order to come to such a conclusion, observers would have to argue that the speed of light must travel faster than normal. This faster speed would be necessary to be able to compensate for the movement of the train during the process of observation.

In other words, because the train is moving in direction x, light would have to travel a little further to hit the mirror on the side of the flatbed car that is nearest to the direction of motion and, thereby, be able to register a light clock tick of time. In order to travel the slightly increased distance due to the motion of the train and still be able to observe a tick rate of once every 6.67×10^{-9} seconds, then, presumably, light would have to travel a little faster than was the case when the clock was stationary.

However, if Einstein is right, and the speed of light remains constant irrespective of the motion of either the source of that light or the motion of an observer with respect to that light, then, what an observer will see when a flatbed car goes by carrying a light clock must be different in some way. More specifically, if the speed of light is the same for everyone independent of motion, then, from the perspective of an observer watching the train go by, light will take longer to reach the mirror on the side of the flatbed car that is nearest to the direction of motion, and, therefore, the rate at which a unit of light clock time will occur will be affected.

The rate of measured time passage for the moving light clock will become longer for such an observer (This is known as time dilation). In other words, from the perspective of an observer watching the train pass by, time will be seen to have slowed down on the moving train.

The rate at which moving clocks in general slow down can be calculated. To make a longer story somewhat shorter, the rate at which a moving clock will be observed to slow down according to the measurements of an observer watching such a moving clock, is represented b: $c/\sqrt{c^2 - v^2}$, where 'c' stands for the speed of light and 'v' gives expression to the speed of the moving body that is carrying a clock. When rearranged, the foregoing expression assumes the following form: $1/\sqrt{1-v^2/c^2}$.

The latter expression is often referred to by the Greek letter: γ, gamma. When 'v' is relatively small compared with the speed of light, 'c' -- which is the case for most things in everyday life -- then gamma will be close to 1, and clocks will appear to run along at nearly the same rates ... that is, although there will be temporal rate differences between moving and stationary clocks, those differences will be so small when 'v' is small relative to 'c' that the time-stretching dimension of moving clocks relative to stationary ones will be very small and difficult to detect.

However, when the speed of 'v' begins to approach the speed of light, gamma's value will start to depart from 1 in a substantial way. For instance, if the speed of 'v' were 90% of the speed of light, then the clock on/in the object that is moving with the speed of 'v' would exhibit a time-lengthening factor that is greater than 2, and, as a result, the moving clock would register the passage of time at a rate that is less than half as fast as a stationary clock.

Conventional interpretations of Einstein's theory of special relativity claim that in the case of objects moving with the aforementioned speed of 'v', then, time is passing at half the rate at which time is measured to pass for a stationary observer. In other words, a relatively stationary observer appears to becoming older at twice the rate as a person riding along on 'v'.

The foregoing considerations are not theoretical speculations. They can be observed to be taking place within an experimental context.

For example, under 'normal' circumstances, the lifetime of a muon lasts approximately 2.2×10^{-6} seconds (2.2 microseconds). However, if one accelerates the speed of muons to nearly the speed of light, something interesting happens.

More specifically, during the latter part of the 1990s, the Brookhaven National Laboratory, located on Long Island in New York State, generated beams of muons that were pushed to nearly the speed of light (99.94 %). This affected how long muons seemed to last.

Given the normal lifespan of a muon (2.2 microseconds), a muon would be able to complete only 15 trips around the 14-foot diameter ring of the Alternating Gradient Synchrotron at Brookhaven. However, when muons were accelerated to nearly the speed of light, they were observed to complete over 400 circuits of the Synchrotron, thereby increasing their usual lifespan by a factor of 29 times, to just over 60 microseconds.

Relativity also has been tested by comparing the readings of atomic clocks located on moving aircraft with readings from the same kind of atomic clock that remain behind on Earth. Richard Keating, an astronomer, and Joseph Hafele, a physicist, conducted this experiment in 1971.

The scientists placed four cesium-beam atomic clocks on commercial aircraft that flew twice around the world, once eastward and once westward. Prior to running the experiment, all the clocks were synchronized with one another.

Upon completion of the flights, the clocks on the planes and the clock at the Naval Observatory were compared with one another. The clocks on the planes all indicated that time was measured to have passed at a slower rate on the planes relative to the measured passage of time for the atomic clock that was located at the United States Naval Observatory.

In addition to the foregoing data, one might take into consideration a form of technology that is used by millions of people every day that only works because it takes into account the principles of relativity – both with respect to the way in which velocity, and gravitational-fields, affect clocks. That technology is known as the Global Satellite Positioning System.

Twenty-four satellites circle the Earth twice a day at an altitude of just over 12,427 miles above our planet. Because those satellites operate at high altitudes, the gravitational field they experience is weaker, and, therefore, their on-board clocks speed up (clocks slow down in stronger gravitational fields and, relatively speaking, speed up in weaker gravitational fields), and, as a result, the GPS satellite clocks gain 45 microseconds a day due to gravitational affects relative to what is happening on the surface of the Earth.

In addition, since the GPS satellites circuit the Earth at approximately 8,699 miles per hour, one also has to factor in the way in which velocity affects clocks. As a result of their orbital velocity, clocks on the GPS satellites slow down daily by 7 microseconds relative to clocks on Earth.

When one combines the effects of gravitation and velocity with respect to the clocks on the satellites and on Earth, the clocks on the satellites speed up a total of 38 seconds a day relative to clocks on Earth. If the foregoing time differential were not taken into consideration, the GPS spatial location readings would be off by an additional 6.2137 miles with each succeeding day.

Consequently, compensating for the foregoing issues requires the clocks on the GPS satellites to be slowed down by 38 microseconds per day. This permits locations on the surface of the Earth to be accurately triangulated.

The foregoing data appears to indicate that people have a choice. (1) They can accept the implications of Maxwell's equations with respect to the constancy of the speed of light and, with Einstein, argue that absolute time and space do not exist because the latter are malleable due to the velocity of moving clocks, as well as due to the strength of the gravitational field in which a clock is located. Or, (2), people can reject Maxwell and Einstein and hold on to their ideas about notions of time and space being absolutes.

Experimental data appears to support the first choice but not the second one. As a result, the idea of time and space being absolutes would seem to be untenable.

However, there might be a third possibility. More specifically, one can acknowledge that, in general terms, relativity is correct and, as a result, one must take into account the impact that velocity and gravitational fields have upon measuring instruments – such as clocks and rulers – nonetheless, accepting the need for such corrections does not necessarily imply that time and space are being deformed in some fashioned.

Einstein once said: "Time is what clocks measure." This is not necessarily true.

Clock measurements (whether in the form of: Light clocks, pendulums, atomic clocks, the decay of particles, the cycles of a cell, or the beating of a heart) serve as indices or markers for what time makes possible. Consequently, clocks are a function of the possibilities inherent in time, but time is not a function of the operational properties of clocks or other kinds of measuring devices.

Yes, the clocks on GPS satellites must be re-calibrated by 38 microseconds each day, but this has nothing to do with the way time passes on satellites relative to Earth. Instead, such alterations have to do only with the manner in which clocks operate when moving and/or when they are subjected to various kinds of gravitational fields.

Moreover, one will notice a difference in the readings for previously synchronized atomic clocks when some of those clocks are put on commercial aircraft and flown around the world and, then, compared with the reading on a clock that has remained stationary on Earth. Once again, however, the foregoing differences merely reflect the way clocks operate when moving or stationary, as well as reflect the way clocks operate in different gravitational fields, and, therefore, those differences do not necessarily reflect changes in either time or space.

Finally, muons traveling at more than 99% of the speed of light will seem to exist for a longer period of time than muons do when those particles are traveling at speeds that are far less than the speed of light. Nevertheless, this is not because the velocity of the muon is affecting time and space more in the first case than in the latter case.

The decay rate of a muon is like a clock. Velocity and/or gravitational fields will affect such decay rates, but changes in that rate really have nothing to do with changes in the nature of time or space.

Dirac was right to introduce relativistic corrections into quantum calculations. Nonetheless, those corrections have to do with the way the process of measurement is affected by differences in velocity and gravitational fields and, therefore, such corrections are not needed to keep track of the way in which the ontological properties of time and space change under various conditions of movement and gravitational fields.

Is time or space malleable? Possibly, but if they are, this is not because the operation of clocks and rulers is affected by movement or sensitivity to the presence of gravitational fields.

Is time travel a possibility? Maybe, but if it is, this is not because clocks moving at different speeds or that are positioned within different kinds of gravitational fields alter the ontological character of time in some fashion.

Is time or space absolute in some sense? Whether, or not, they are, measurement will not disclose the truth of the matter.

When someone says that moving bodies experience time dilation – that is, a stretching out of the passage of time – one wonders about the dynamics of that process. Exactly how does velocity engage and alter the ontology of time? Or how does a gravitational field engage and alter the ontology of time?

Physicists cannot answer either of the foregoing questions. All statements about the passage of time are a function of some kind of measurement process.

Perceptions and interpretations are being filtered through the properties of one, or another, system of measurement. Therefore, what one 'sees' or understands in this regard tends to reflect the properties of the measuring process itself rather than the properties of time.

To be sure, we know that something called "time" is needed in order for movement to be able to take place. Nonetheless, the facticity of movement (or the measurement thereof) tells us nothing more about the ontological character of time other than that the former is, in some way, functionally dependent on the latter.

Galileo might have been correct when he indicated that human beings are not able to detect (empirically) the presence of absolute motion. However, one could agree with Galileo (as Einstein did) without simultaneously being logically required to admit there is no such thing as absolute motion (and I have no idea whether absolute motion exists or what this would even mean).

The problem of detection is a methodological issue. The issue of whether, or not, absolute motion exists is an ontological matter, and Galileo's foregoing claim is really only an acknowledgement that the nature and limits of methodology will always interfere with our capacity to determine whether, or not, such a thing as absolute motion exists.

Similarly, Einstein might have been correct when he indicated that the process of measurement generates variable results when bodies are in motion relative to one another or when those bodies encounter gravitational fields of different kinds. Nonetheless, even if something called "absolute time" and "absolute space" existed (and, again, I have no idea whether absolute time and space exist or what even would be meant by such terms), all that follows from Einstein's perspective is the idea that measurement is a malleable process, and, consequently, such malleability would always interfere with our capacity to be able to detect the presence of absolute time and space … if they existed in some sense.

Galileo and Einstein – each in his own way -- were forerunners to the indeterminacy issue first introduced formally by Heisenberg (and not what that idea became through the hermeneutical influence of Bohr). Measurement often gets between, on the one hand, researchers/observers and, on the other hand, the actual nature of the reality that is being filtered through such measurement processes.

For Einstein, the principle of special relativity was the conceptual glue that tied physical phenomena together. Special relativity ensured that all physical phenomena obeyed the same set of laws irrespective of the motion of the bodies being considered … that is, the laws of nature are independent of the motion which different observers have with respect to one another.

Essentially, the special theory of relativity is a way of demonstrating that the laws of nature are invariant and independent of any given frame of reference. The Lorentz transformation equations provide a means of establishing the equivalencies of the measurement perspectives of different frames of reference and, in the process, show that while any given process of measurement (involving clocks and rulers) might vary from framework to framework due to the effects of motion, nonetheless, one could translate those differences in measurement from one framework to another, and the laws of nature would prove to be invariant across those translational processes involving measurement.

When someone asked Einstein a question about the meaning of the theory of relativity, Einstein is reported to have responded in the following way. He indicated that the theory of relativity was sort of like when one sits on a hot stove for a minute, it seems like an hour, but when one sits next to a beautiful woman for an hour, it seems like a minute.

He might have had his tongue firmly in his cheek when he related the foregoing similes. However, people have been misled by the foregoing imagery.

Despite its terminology, special relativity does not give priority to the issue of relativity. Instead, special relativity gives primary emphasis to the invariance of the physical laws that tie frameworks together that are in relative motion with respect to one another ... an invariance that can be demonstrated by showing how the measurement processes used in such frameworks can be translated in ways that preserve the laws of nature.

Under different conditions of motion, clocks might run faster or slower. Nonetheless, irrespective of the variability displayed by such clocks due to the effects of motion on the operation of those clocks, the physical laws governing the dynamics of a system remained the same quite independently of the framework through which one observed those dynamics.

All manner of speculative gibberish concerning time travel have been read into the meaning of Einstein's theory of special relativity by novelists, science fiction movies, television shows, and the popular press. However, what is truly special about that theory is the manner in which it demonstrates how the physical laws of dynamics are conserved despite the malleability of the measurement process under different conditions of motion.

Chapter 8: Searching For Unity

Prior to Newton, scientists believed that the effects of gravity manifested themselves in two, distinct ways. One modality involved the manner in which gravity affected phenomena on Earth, while the second modality concerned the manner in which gravity affected celestial phenomena.

In 1687 Newton showed that the gravitational phenomena of both Earth and the heavens were expressions of one underlying force. While Newton suspected, but was unable to prove, that there were other forces at work in nature, his universal theory of gravitation was an early step in the attempt to demonstrate that the complexity exhibited by phenomenal events could be treated as a function of the way in which some underlying, singular force variably manifested itself under different conditions.

Nearly two hundred years later (minus 20 years or so), James Maxwell was able to show that electricity and magnetism were not separate phenomena. Instead, he successfully argued they were different manifestations of one underlying force that manifested itself in variable ways depending on circumstances.

More than a half century later – and despite early successes arising from his special and general theories of relativity, as well as due to his insights into the significance of Brownian motion for atomic theory, together with the implications of the photoelectric effect for quantum theory -- Einstein ran into a brick wall when it came to uncovering the nature of the deep unity he felt was operative at the heart of physical phenomena. As a result, Einstein spent the last several decades of his life trying – unsuccessfully -- to develop a unified field theory that would be capable of showing that the phenomena of gravity and electro-magnetism were manifestations of one underlying force.

Beginning in 1967, twelve years after Einstein passed away, Steven Weinberg, Abdus Salaam, and Sheldon Glashow independently laid the foundations for a theory that indicated how the weak force and the electromagnetic force might be different manifestations of one underlying phenomenon that arose after some form of spontaneous symmetry breaking had occurred, and this became known as the electroweak model. Seven years later, in 1973, the first confirmation of their theory came in the form of the discovery of the existence of neutral currents in neutrino scattering experiments, and sixteen years later, in 1983, a second confirmation of their theory came via the discovery of W and Z particles at CERN.

Many physicists believe that gravity, electromagnetism, the strong force, and the weak force are expressions of some underlying, singular phenomenon that begins to manifest itself as four different forces following some kind of spontaneous symmetry breaking that occurs ... perhaps in conjunction with the Big Bang. Nonetheless, uncovering proof concerning the existence of such a fundamental unification of forces has been elusive.

If one leaves the force of gravitation out of the picture, for the moment, there have been a variety of attempts to unite electromagnetism, the weak force, and the strong force in the form of what are known as grand unified theories (GUT). Such theories predict that at very high energies – above $10^{15\text{-}16}$ GeV – strong, weak, and electromagnetic forces might exhibit the same kind of strengths.

Above some given level of energy – known as the grand unification scale – such theories indicate that just one kind of force will be present. However, below that threshold value, and following some form of spontaneous symmetry breaking, three forces will manifest themselves and give expression to phenomena involving different strengths and properties that are in accord with what has been observed in modern physics.

The most powerful collider in existence today is the LHC facility at CERN in Switzerland. This operates at energies that are many magnitudes of order lower than the aforementioned $10^{15\text{-}16}$ GeV, and, therefore, in the foreseeable future – if ever -- grand unified theories cannot be probed directly, but there still are ways of testing such models.

For example, grand unified theories predict that quarks can change into leptons (e.g., electron, muon and tau particles along with their respective neutrinos). Another way of saying the same thing is to predict that protons (made from three quarks) will decay into some assortment of lepton debris.

More specifically, Grand Unified Models indicate that the lifespan of a proton should be about 10^{32} years. A variety of scientific projects in different countries have tested the foregoing prediction and, to date no protons have been observed to decay ... and one should note that such experiments have been taking place, in one form or another, for several decades.

Based on the foregoing findings, one cannot necessarily claim that protons do not decay. However, what those results do indicate is that if protons decay, then that process would have to occur at some temporal juncture beyond the predicted lifespan of 10^{32} years.

If one fails to find evidence that protons decay within 10^{32} years, then, one can, of course, keep increasing the proposed lifespan of a proton to some longer period of time. At some point, however, one must face the question of whether, or not, protons really do decay at all.

Subsequent, experimental data might, yet, confirm that protons do decay. Nonetheless, at the present time, what we know is that the original GUT prediction concerning the estimated lifespan of a proton appears to be incorrect.

There also are various theoretical considerations that have led some physicists to believe that the strengths of the three aforementioned forces (strong, weak, and electromagnetic) will not be exact at the unification scale of energies unless certain kinds of conditions exist. One set of such conditions is given expression through what are known as Supersymmetry theories.

Symmetry exists in a given physical context, when one can perform various kinds of transformational operations in conjunction with that context, and, yet, the laws governing such a context do not change as a result of those operations. The symmetry involved in Supersymmetry models has to do with the relationship between fermions (matter particles such as quarks and leptons) and bosons (force mediating or carrying particles).

Supersymmetry claims that fermions and bosons are not really distinct entities, but, rather, under the right circumstances, they are transformable into one another. Fermions are particles with a ½ spin value, whereas bosons have an integer spin value.

Just as Paul Dirac argued some 85 years ago that when one brings relativity and quantum theory together, the spacetime symmetries entailed by such a co-joining of ideas implies the existence of antiparticles, so too, Supersymmetry posits the existence of a quantum variation on spacetime – known as superspace – in which particle symmetries exist that link fermions and bosons together. Superspace is not so much about the notion of space in any usual sense (e.g., up-down, right-left), as it is about a physical context that gives expression to the presence of certain conditions involving fermionic possibilities.

A quantized 'movement' or transition dynamic in such a context means that a particle (whether fermion or boson) assumes a given fermionic possibility in superspace. If a boson makes the foregoing kind of transition movement in superspace, then, that boson becomes a particular kind of fermion as a function of whatever fermionic possibility the boson assumes during such a dynamic, quantized transition, whereas, on the other hand, if a fermion undergoes a transition movement within superspace, it becomes a boson in accordance with the properties it assumes in superspace.

Supersymmetric models that are referred to as "natural" (and, these are considered to be the simplest and most heuristically powerful Supersymmetric models) involve properties that provide a means of suppressing the activities of virtual particles, and, thereby, avoid the theoretical problems that arise when virtual particles are permitted to interact in unrestricted ways with non-virtual particles. If a model lacks the properties necessary to satisfy the conditions for being naturally Supersymmetric, then some physicists believe that these kinds of models must contain some other means for placing limits on the extent to which virtual particles can interact with particles if, among other things, the vacuum energy problems being alluded to are to be avoided.

If Supersymmetry is correct, then, given the right conditions, the spin values of fermions and bosons can be changed back and forth. Among other things, this provides a way to get rid of the infinities that tend to surface in calculations involving high-energy particle dynamics.

Supersymmetry also predicts the existence of a spin-2 particle. This particle is thought to be consistent with the graviton -- the carrier or mediator of gravitational force – and, therefore, those kinds of theories are sometimes referred to as supergravity theories.

Initially, GUT models sought to unify three forces – strong, weak, and electromagnetic. However, when theoretical modifications are made to include the idea of Supersymmetry (SUSY), then, as indicated in the previous paragraph, gravity enters the picture as well, and, as a result, such a modified theory gives expression to an attempt to unify four forces rather than just three forces.

The theoretical structure of Supersymmetry indicates that every fermion and every boson must have a Supersymmetric counterpart. Consequently, there are twice as many particles in Supersymmetry models as in the Standard Model of physics.

Bosons have fermionic partners. Fermions have bosonic counterparts.

Thus, bosons such as photons and gluons are predicted to have Supersymmetric fermionic partners known as photinos and gluinos respectively. On the other hand, fermions such as electrons and quarks will have bosonic Supersymmetric partners known as selectrons and squarks.

Supersymmetric fermionic partners have a suffix of –ino that replaces the last part of the name for any given boson particle. Supersymmetric bosonic partners have a prefix of 's-' added to the usual name of a given fermion.

The Supersymmetric fermionic partners are what bosons are transformed into when the latter's spin value is changed from an integral into a fractional value. The Supersymmetric bosonic partners are what fermions are transformed into when the spin value of the latter particles is changed from a fractional value into an integral value.

Just as various attempts at unification in physics have sought to reduce – under the right circumstances -- the number of known forces down to one fundamental force, a similar kind of effort has been made with respect to particles. In Supersymmetry models, instead of having two classes of particles – namely, fermions (matter particles such as quarks and leptons) and bosons (force carrying or mediating particles such as photons and gravitons), there exists only one class of Supersymmetric particles that, given the right conditions, can be manifested in the form of fermions or bosons.

Supersymmetric models predict the existence of an array of particles (i.e., the Supersymmetric partners to fermions and bosons alluded to earlier) that are in addition to the particles contained within the Standard Model. To date, none of the predicted particle partners have been discovered.

The experiments that currently are getting under way through CERN in Switzerland might generate data that confirms the existence of some, or all, of the foregoing sorts of particles. Nonetheless, some physicists believe that at least some of the missing Supersymmetric particles are likely to have masses that are sufficiently small and, consequently, should have shown up within the energies at which the LHC has been operating, and, therefore, the fact that particles with those sorts of masses have not, yet, been detected is potentially troublesome.

Despite the foregoing source of anxiety amongst some physicists, the upside of Supersymmetric models is considerable, and, as a result, many physicists are not ready to abandon those theories despite the absence of evidence for the existence of the superpartner particles. Indeed, if true, the right kind of Supersymmetric model could account for why different forces have the strength they do as well as be able to explain why various particles have the masses they do.

In addition, Supersymmetric models have the potential to be able to solve the vacuum energy problem discussed in the last chapter, as well as to provide a means of avoiding many of the infinities that tend to arise in conjunction with calculations involving high-energy particle physics. Moreover, some physicists even believe that such models could provide a solution to the mystery of dark matter.

Finally, as pointed out in the chapter on string theory, Ed Whitten has revealed the existence of dualities that establish the equivalency between, on the one hand, a ten-dimension version of superstrings and, on the other hand, supergravity (which is a Supersymmetric model that contains gravitational properties). Consequently, if the right version of Supersymmetry turns out to be true, then, a great deal of mathematical wherewithal might already be present to help to further develop those kinds of models, and, in fact, as a result of the aforementioned dualities, Supersymmetry could be one of the doors of opportunity through which string theory finally comes into its own and begins to pay some of the theoretical and practical dividends that have been promised for so long by the string theory lobby.

Quantum Queries

Nonetheless, despite the foregoing sorts of possibilities, Supersymmetric models are not without their problems. For example, discovering that the mass of the Higgs particle (or, at least, one of them) comes in at around 125 GeV already has eliminated some of the best Supersymmetric theories from the running.

More specifically, in those Supersymmetric models that are referred to as being "natural", the masses of the superpartners should not be that much heavier than the mass of the Higgs particle. Yet, even though a Higgs particle has been found, nevertheless, no superpartner particles that are near to the mass of the Higgs have been detected so far.

If superpartners exist but are heavier than what is predicted by natural Supersymmetric models, then, theorists stand to lose an attractive feature of so-called natural versions of Supersymmetry – namely, their capacity to suppress the way in which virtual particles interact with non-virtual particles and, in the process, help eliminate the presence of infinities in calculations involving those kinds of interactional dynamics. While theorists still might come up some other kind of non-'natural' mechanism to account for the important suppression property touched on previously, such alternative mechanisms are not likely to be as simple as the ones that appear in natural Supersymmetric models.

Because of their capacity to account for the suppression of virtual particle activity in conjunction with non-virtual particles, natural Supersymmetric models also possess a stabilizing effect with respect to the vacuum. However, if Supersymmetric models are jettisoned from the picture, and if some other kind of suppression mechanism is not forthcoming, then, many physicists believe that the stability of the vacuum will depend on the mass of the Higgs particle.

The heavier the mass of the Higgs is, then, the more stable is the vacuum likely to be. The mass for the Higgs-like particle that was announced in 2012 is at the lower end of the mass scale

| Quantum Queries |

required to lend full stability to the vacuum, and, therefore, under such circumstances, the vacuum could persist as a stable medium for a very long time, and, yet, ultimately still could prove to be susceptible to instabilities at some point.

The foregoing condition is known as a "metastable state." According to some physicists, quantum fluctuations of the right kind could destabilize such a vacuum and, thereby, set in motion a catastrophic, cascading set of events that could bring an end to the universe in its present form.

The foregoing several hundred pages have outlined some of the possibilities and problems associated with certain dimensions of physics ... especially in relation to quantum physics and particle physics. While far from complete, the aforementioned overview does provide, I believe, a fairly good – if limited – account of what modern physics has to offer as a way to engage the reality problem that constitutes one of the two central themes that form the woof and warp of the explorations that begin with Volume I of *Final Jeopardy*.

The second central theme alluded to above revolves about the challenge of 'Final Jeopardy.' This latter challenge involves the problem of trying to figure out how to assess the heuristic value of any given approach to the reality problem as far as the issue of how best to proceed through life is concerned.

On the asset side of the ledger, modern physics has introduced human beings to, among other things, ideas involving: Constants, antimatter asymmetries, entanglement phenomena, the origins of mass, quantum dynamics, special relativity, and so-called 'theories of everything'. On the liability side of the same ledger, modern physics has not, yet, been able to close the deal, so to speak, with respect to resolving many of the problems and questions that continue to plague the foregoing asset entries.

For instance, physicists don't know why constants have the values they do or how such constants came to have those particular values. In addition, while physicists acknowledge the existence of antimatter asymmetries and entanglement phenomena in the universe, at the present time, those individuals appear to have a very limited understanding of why such asymmetries exist or what the full scope of the entanglement phenomenon involves.

Furthermore, a great deal has been written about various candidates for a theory of everything. Nonetheless, each of those candidates is riddled with many questions and problems … questions and problems that might, or might not, be answered in the foreseeable future … if at all.

Quantum dynamics has introduced all manner of weirdness into the conversation (e.g., inherent indeterminacy and randomness, cats that, thanks to the superposition principle, supposedly are both alive and dead, spooky action at a distance, probability distributions that are ontological in nature and not just methodological in character). However, quite possibly, such weirdness might be more a function of speculative hermeneutics than it actually reflects the nature of reality.

The accomplishments of quantum physics are quite remarkable without the weirdness sideshow. Indeed, the various dimensions of weirdness in quantum physics seem to be more about trying to allay the discomfort some physicists feel in relation to the fact that while quantum theory leads to correct answers, oftentimes physicists are not quite sure what is going on beneath that surface precision.

For example, physicists use the term "quantum jitters". This is a phrase that refers to the inherent fluctuations that are allegedly entailed by quantum dynamics, and, yet, the only jitters that are present might belong to the physicists who develop hermeneutical perspectives to account for something that could be just a projection of their own theoretical insecurities onto reality.

In an earlier chapter, I indicated that physicists might not have a good model for the vacuum. In addition, comments have been made earlier about the possibility that the way in which physicists think about the notion of virtual particles could also be flawed.

In both cases, when one engages reality through the filters of quantum physics, one encounters problems of infinities of one kind or another. Those infinities might be more a function of the theory being used rather than a reflection of the nature of reality.

One of the attractive features of natural Supersymmetric theories is their capacity to suppress the way in which virtual particles supposedly interact with non-virtual particles, and, in the process, avoid various kinds of problems involving infinity. This feature is so important that physicists believe that if natural Supersymmetric models turn out to be incorrect (and, to some extent, some of the data from the LHC at CERN seems to be pointing in this direction), then, nonetheless, many physicists believe that some alternative method must be developed as a means of suppressing the way in which virtual particles and non-virtual particles supposedly interact.

Yet, if physicists are operating through a flawed understanding concerning the nature of vacuum energy, as well as in relation to the dynamic properties of virtual particles, then, the foregoing feature might become unnecessary. In other words, some physicists might be suffering from an iatrogenic-like disease in which the very methodology that is brought to bear on exploring reality is the source of at least some of the theoretical maladies with which physicists are trying to contend through, among other models, Supersymmetry.

When considering the foregoing possibilities, one should keep something in mind. I am not saying there is not some kind of energy associated with the vacuum, nor am I saying that something like virtual particles (the reality might be otherwise) aren't at work in conjunction with non-virtual particles, but, instead, what is being said is that current ideas concerning the nature of the vacuum energy and the nature of virtual particles are problematic (e.g., the infinity issue, along with the observed difference between prediction and actual measurements in relation to such ideas).

The energy that is present in the vacuum might not be a function of the inherent jitteriness of quantum dynamics. Instead, such energy might be a function of the local dynamics involving a variety of fields – including the Higgs field, the gravitational field, the electromagnetic field, and so on.

Similarly, to whatever extent something like virtual particles exist, they might be giving expression to local conditions. As such, virtual particles might only arise under a limited set of conditions that operate in accordance with the dynamics that are taking place under those local conditions rather than being a function of a theoretical perspective that is being globally imposed upon virtually every point in space and, in the process, laying the groundwork for subsequent problems with infinities.

Another aspect of weirdness involving modern physics involves special relativity. All manner of imaginative ideas have been read into the possible implications of the special theory of relativity with respect to issues that involve time travel.

Yet, when one critically examines the dynamics of special relativity, one discovers that many people – including physicists – might not only have been conflating the malleability of measurement under different conditions of motion and gravitational effects with a very different issue – namely, the malleability of time. In addition, many of those same physicists seem to have forgotten that special relativity is primarily about preserving the invariance of physical laws when engaged through different frameworks of relative motion rather than being about the relativity of time and space as ontological entities.

When one removes all the speculative smoke and mirrors from modern physics, the latter still gives expression to a very impressive set of accomplishments. Nonetheless, the model of reality that remains after its mirror has been wiped clean of the foregoing sort of hermeneutical smoke residue continues to harbor many unanswered questions, unresolved problems, as well as areas that are woefully incomplete, and, therefore, the resulting model blurs and distorts the nature of reality in a variety of ways.

Notwithstanding the foregoing considerations, let's entertain a thought experiment of sorts. Imagine that some kind of string theory, or brane theory, or M-theory, or Supersymmetric model of reality arose that permitted all of the outstanding questions and problems in quantum and particle physics to be definitively answered or resolved, and, in the process, give expression to some sort of 'theory of everything.'

What would the possible significance of such a theory be with respect to the 'Final Jeopardy' issue and the reality problem? Perhaps, the first order of business would be to ask whether such a development actually constitutes a 'theory of everything'?

While many physicists are of the opinion that the laws of physics govern the way the entire universe operates, this is not necessarily so. Even if one had a complete theory of everything as far as the physical universe is concerned, one would not necessarily be able to proceed and tenably conclude that: consciousness, reason, understanding, creativity, language, the self, morality, emotional intelligence, and spirituality are all functions of the physicists' version of a theory of everything.

In fact, it is quite possible that the physicist's version of a theory of everything accounts for only very limited dimensions among the possibilities that are inherent in the universe. Physics might be the key to solving many problems concerning the physical universe, and, as well, physics might be the royal road to exploiting various aspects of the physical universe (e.g., in the form of electronics, chemistry, materials science, and other forms of technological innovation), and, yet, there is a very real possibility that even a complete theory of everything from the perspective of physicists would not be able to answer any of the questions that are most important to human beings.

What is the nature of human potential, and how did the various dimensions of that potential arise? What makes logic, reason, consciousness, critical thought, understanding, and memory possible?

What, if anything, does morality and spirituality have to do with the nature of the universe? What does it mean to be a sovereign human being and how is such sovereignty to be reconciled with the notion of community?

What are the origins of creativity and inventiveness? How did the capacity to communicate arise in human beings?

The current state of physics cannot provide a plausible response in relation to any of the foregoing questions. Furthermore, there is no guarantee that even a complete physical theory of everything could do so either.

Like Laplace, many physicists believe they have no need for any hypothesis that resides beyond the horizons of physics. Nevertheless, such physicists cannot provide even the most rudimentary account of all the capacities – in the form of awareness, logic, reason, creativity, imagination, language, insight, understanding, morality, and communication – that makes the practice of physics possible.

Even if one were to put the foregoing considerations aside, physics entails a very real problem as far as the Final Jeopardy issue is concerned … a problem that was touched upon in Chapter 7 toward the end of the discussion concerning the size of the proton. More specifically, even if one: Were born in 1880; hit the conceptual ground running in relation to the contributions of Planck, Einstein, Bohr, Heisenberg, Schrödinger, Born, Jordan, Pauli, De Broglie, Dirac, along with so

| Quantum Queries |

many others; understood everything they said, and lived for 135 years with full use of all one's faculties, nonetheless, except in limited ways, one still would not know today what the nature of reality is and one would not necessarily have any deeper insight into how to respond to the Final Jeopardy challenge than anyone else did who was not a physicist.

No one knows where things will stand in physics a hundred years from now, but for people like me who are coming to the end of their lives, physics over the last hundred-plus years would not have been of much assistance with respect to resolving the Final Jeopardy issue. To be sure, physics could have provided some degree of insight into the reality problem, and, as a result, it might have been able to shed some degree of light on how – within certain parameters -- one might go about engaging the Final Jeopardy challenge, but, nonetheless, on the whole and to date, physics has not been able to rigorously address any of the questions and problems that seem to be at the heart of most every human being on Earth.

To acknowledge the foregoing point, does not denigrate either physics or physicists. Rather, the foregoing comments are intended to help put things in perspective.

As incredibly ingenious, imaginative, insightful, and heuristically powerful sciences like physics might be, they have limits with respect to what they can accomplish in the near term. Yet, human beings live in the near term, and not in the long term.

Whatever science might be able to accomplish over the next several hundred years, the only thing of relevance it has to offer for most human beings is what can it do for such individuals now with respect to the issues that are of current concern to the vast majority of human beings. What can it explain now? What can it account for now? What knowledge about the nature of reality does it have to offer now? How can it – or can it – assist human beings to critically engage the Final Jeopardy challenge … namely, how to use the time available to one most effectively as far as engaging the basic questions of life are concerned – all of which are variations on the reality problem.

When I was examined in conjunction with defense of my doctoral dissertation (and there was a physicist and a biophysicist on the examination committee), several of my examiners wanted to know whether what I had done in the thesis would be of any value to other human beings, or even whether, or not, anyone would be interested in what I had to say. While I understood why they were asking such questions, the individuals who raised those sorts of issues seemed to miss something of considerable importance.

To be sure, defending my dissertation meant that I had to do so in a way that would satisfy the concerns of the examiners with respect to their expectations about what a doctoral candidate should be able to demonstrate in the way of an understanding concerning his or her alleged area of expertise. Nevertheless, the primary reason why I wrote the dissertation was as a means of clarifying to myself various issues concerning the reality problem and how I might best proceed with respect to engaging the Final Jeopardy challenge ... and a dissertation process that could not advance such purposes was of limited use to me irrespective of whatever purposes might have been served by the dissertation process in general as understood from the perspective of some of my examiners.

A similar point can be made in relation to the practice of physics. However valuable the activity of physics might be from any number of perspectives, and however intriguing and fulfilling such activities might be for those who are immersed in them, the value that such activities have for me – and for many other individuals -- is a function of whether, or not, those activities can shed light on the problems and issues that are of most importance to me and to many other human beings ... which, to a considerable degree, physics cannot do.

In many ways, physicists are in a position that is very similar (although not exactly so) to the guy in the Chinese Room

| Quantum Queries |

251

Problem (cf. John Searle) that was explored in the third chapter of the first volume of *Final Jeopardy*. Physicists exist in their own scientific room where questions are passed to them from the mysterious realm beyond (i.e., reality problem) by being slid beneath the door that separates physicists from the rest of the world.

Physicists have been provided with a book of algorithms (i.e., formulas, equations, mathematical expressions, field theories, ideas about symmetry, etc) that are used to arrange an array of theoretical symbols for purposes of describing the nature of an unknown reality that is being asked about through the questions that are slipped into the room through the crack beneath the door that separates the two realms. Physicists have become very proficient with respect to arranging an array of symbols in accordance with the algorithms that are listed in their book of instructions about how to do physics, and, as a result, they are able to answer, in very precise ways, all manner of questions that are addressed to them from the other side of the door.

Nonetheless, despite the foregoing kind of competency, one still could query the nature of just what it is that the physicists actually understand about the nature of reality. On the one hand, they have facility with a language – namely, mathematics – and, yet, on the other hand, understanding how to work successfully with that language doesn't necessarily guarantee that physicists understand the nature of the reality being described through that language ... any more than the guy in the Chinese Room Problem understands the Chinese language despite being able to answer all manner of questions in Chinese as a result of gaining proficiency with the book of algorithms with which he has been provided in order to be able to arrange symbols in ways that appear to respond appropriately to the questions that are being passed to him from the next room.

In 30-40 years (maybe even sooner), someone might have to critically reassess the whole issue of physics and the reality problem in the context of the Final Jeopardy challenge. At the present time, however, there seem to be many reasons (some of which have been stated previously) for coming to the conclusion that despite all the media hoopla surrounding the accomplishments of physicists, nonetheless, what physicists have succeeded in achieving appears to have very limited value with respect to both the reality problem as well as the Final Jeopardy issue.

To be sure, one should take note of what physics has to offer. For example, among other things, what is discovered through physics can help eliminate any number of theories concerning the nature of reality, and this is valuable information to have when considering various possibilities about how to engage the Final Jeopardy issue

On the other hand, one also should keep in mind that at the present time, physics is limited, incomplete, and problematic in a variety of ways (some of which have been explored throughout this book). Therefore, as far as being able to provide much insight into how to handle the challenge of Final Jeopardy, physics seems over-matched ... and this state of affairs might continue to be true even if someone were to come up with a defensible theory of everything as far as physical phenomena are concerned.

Bibliography

Articles

Zeeya Mer Ali, 'Gravity Off The Grid', pp. 44-51, *Discover*, March 2012.

Ross D. Andersen, 'An Ear To The Big Bang', pp. 40-47, *Scientific American*, October 2013.

Zvi Bern, Lance J. Dixon and David Kosower, 'Loops, Trees and the Search for New Physics', pp. 34-41, *Scientific American*, May 2012.

Jan Bernauer and Randolf Pohl, 'The Proton Radius Problem', pp. 32-39, *Scientific American*, February 2014.

Yudhijit Bhattacharjee, 'Paranormal Psychologist', pp. 52-58, *Discover*, March 2012.

Leo Blitz, 'The Dark Side of the Milky Way', pp. 36-45, *Scientific American*, October 2011.

Deborah Blum, 'The Scent of Your Thoughts', pp. 54-57, *Scientific American*, October 2011.

Alan Burdick, 'The Sixth Sense: Time', pp. 8-11, *Discover Magazine – The Brain*, Spring 2011.

Peter Byrne, "The Many Worlds of Hugh Everett", *Scientific American*, October 21, 2008.

David Castlevechhi, 'Is Supersymmetry Dead?' pp. 16-18, *Scientific American*, May 2012.

Timothy Clifton and Pedro G. Ferreira, 'Does Dark Energy Really Exist?', pp. 48-55, *Scientific American*, April 2009.

Tamara Davis, 'Is the Universe Leaking Energy', pp. 38-47, *Scientific American*, July 2010.

Karl Deisseroth, 'Conrolling the Brain With Light', pp. 48-55, *Scientific American*, November 2010.

Cara Feinberg, 'The Placebo Phenomenon', pp. 36-39, *Harvard Magazine*, January-February 2013.

Jonathan Feng and Mark Trodden, 'Dark Worlds', pp. 38-45, *Scientific American*, November 2010.

Douglas Finkbeiner, Meng Su, and Dmitry Malyshev, 'Giant Bubbles of the Milky Way', pp. 42-47, *Scientific American*, July 2014.

Tim Folger, 'Second Genesis', pp. 18-24, *Discover Magazine – Extreme Universe*, Winter 2010.

Tim Folger, 'How Can You Be In Two Places At Once?', pp. 56-61, *Discover Magazine – Extreme Universe*, Winter 2010.

Avishay Gal-Yam, 'Super Supernova', pp. 44-49, *Scientific American*, June 2012.

Donald Goldsmith, 'The Far, Far Future of Stars', pp. 32-39, *Scientific American*, March 2012.

Andrew Grant, 'Night Ranger', pp. 32-35, *Discover Magazine – Extreme Universe*, Winter 2010.

Andrew Grant, 'Enter String Man', pp. 70-73, *Discover Magazine – Extreme Universe*, Winter 2010.

Trisha Gura, 'When Pretending Is The Remedy', pp. 34-39, *Scientific American Mind,* March/April 2013.

Martin Hirsch, Heinrich Pas, and Werner Porod, 'Ghostly Beacons of New Physics', pp. 40-47, *Scientific American*, April 2013.

Ray Jayawardhana, ' Coming Soon: A Supernova Near You', pp. 68-73, *Scientific American*, December 2013.

Meinard Kuhlmann, 'What Is Real?', pp. 40-47, *Scientific American*, August 2013.

Robert Kunzig, 'Sticky Stuff', pp. 50-55, *Discover Magazine – Extreme Universe*, Winter 2010.

Robert Kunzig, 'The Unbearable Lightness of Neutrinos', pp. 62-69, *Discover Magazine – Extreme Universe*, Winter 2010.

Michael D. Lemonick, 'The Dawn of Distant Skies', pp. 40-47, *Scientific American*, July 2013.

Michael D. Lemonick, 'Big Bang', pp. 12-17, *Discover Magazine – Extreme Universe*, Winter 2010.

Noam I. Libeskind, 'Dwarf Galaxies and the Dark Web', pp. 46-51, *Scientific American*, March 2014.

Don Lincoln, 'The Inner Life of Quarks', pp. 36-43, *Scientific American*, November 2012.

Joseph Lykken and Maria Spiropulu, 'Supersymmetry and the Crisis in Physics', pp. 34-39, *Scientific American* 2014.

Michael Moyer, 'Is Space Digital', pp. 30-36, *Scientific American*, February 2012.

Steve Nadis, 'First Light', pp. 38-45, *Discover*, April 2014.

Corey Powell, 'When a Slumbering Monster Awakes', pp. 62-63, *Discover*, April 2014.

Heather Pringle, 'The Origins of Creativity', pp. 36-43, *Scientific American*, March 2013.

Marcus E. Raichle, 'The Brain's Dark Energy', pp. 44-49, *Scientific American*, March 2010.

V.S. Ramachandran, 'True Vision', pp. 32-40, *Discover Magazine – The Brain*, Spring 2011.

Michael Riordan, Guido Tonelli and Sau Lan Wu, 'The Higgs At Last', pp. 66-73, Scientific American, October 2012.

Subir Sachdev, 'Strange and Stringy', pp. 44-51, *Scientific American*, January 2013.

Eric Scerri, 'Cracks in the Periodic Table', pp. 68-73, *Scientific American*, June 2013.

Caleb Scharf, 'The Benevolence of Black Holes', pp. 34-39, *Scientific American,* August 2012.

Steven Stahler, 'The Inner Life of Star Clusters', pp. 44-51, *Scientific American*, March 2013.

Paul J. Steinhardt, 'The Inflation Debate', pp. 36-43, *Scientific American*, April 2011.

Vlatko Vedral, 'Living in a Quantum World', pp. 38-43, *Scientific American*, June 2011.

Hans Christian von Baeyer, 'Quantum Weirdness?' pp. 46-51, *Scientific American*, June 2013.

Carl Zimmer, 'The Surprising Origins of Life's Complexity', pp. 84-89, *Scientific American*, August 2013.

Books

Lyndon Ashmore, *Big Bang Blasted!:The Story of the Expanding Universe and How It Was Shown to be Wrong*, Book Surge, 2006.

Halton Arp, *Seeing Red: Redshifts, Cosmology and Academic Science*, Apeiron, 1998.

Peter Atkins, *Four Laws: What Drives the Universe*, Oxford University Press, 2007.

Ian G. Barbour, *Myths, Models and Paradigms: A Comparative Study In Science and Religion*, Harper & Row Publishers, 1974.

John D. Barrow, *New Theories of Everything*, Oxford University Press, 2007.

John D. Barrow, *The Constants: From Alpha to Omega – The Numbers That Encode the Deepest Secrets of the Universe*, Random House, 2002.

Susan Blackmore, *Consciousness: An Introduction*, Oxford University Press, 2004.

David Bohm, *Wholeness and the Implicate Order*, Ark Paperbacks, 1983.

Harold I. Brown, *Perception, Theory and Commitment: The New Philosophy of Science*, The University of Chicago, 1977.

Brian Clegg, *Before the Big Bang: The Prehistory of Our Universe*, St. Martin's Press, 2009.

Brian Clegg, *The God Effect: Quantum Entanglement, Science's Strangest Phenomenon*, St. Martin's Press, 2006.

Frank Close, *The Infinity Puzzle: Quantum Field Theory and the Hunt for an Orderly Universe*, Basic Books, 2011.

Frank Close, *Antimatter*, Oxford University Press, 2009.

Brian Cox & Jeff Forshaw, *Why Does E=mc²?*, Da Capo Press, 2009.

Robert P. Crease and Charles Mann, *The Second Creation: Makers of the Revolution in 20th-Century Physics,* Collier Books, 1986.

Paul Davies, *Cosmic Jackpot: Why our Universe Is Just Right For Life*, Houghton Mifflin, 2007.

Guy Deutscher, *Through the Language Glass: Why the World Looks Different in Other Languages*, Metropolitan Books, 2010.

A.C. Ewing, *A Short Commentary on Kant's Critique of Pure Reason*, The University of Chicago Press, 1938.

Harald Fritzsch (translated by Gregory Stodolsky), *The Fundamental Constants: A Mystery of Physics*, World Scientific Publishing Company, 2009.

Louisa Gilder, *The Age of Entanglement: When Quantum Physics was Reborn*, Alfred A. Knopf, 2008.

Malcolm Gladwell, *Blink: The Power of Thinking Without Thinking*, Little, Brown and Company, 2005.

Peter Godfrey-Smith, *Theory and Reality: An Introduction to the Philosophy of Science*, University of Chicago Press, 2003.

Rebecca Goldstein, *Incompleteness,* W.W. Norton & Company, 2005.

Nelson Goodman, *Ways of Worldmaking*, Hackett Publishing Company, 1978.

Stephen Hawking, *A Brief History of Time: From the Big Bang to Black Holes*, Bantam Books, 1990.

Nick Herbert, *Elemental Mind: Human Consciousness and the New Physics*, Dutton, 1993.

Nick Herbert, *Quantum Reality: Beyond the New Physics*, Anchor Press/Doubleday, 1985.

John Holland, *Emergence: From Chaos to Order*, Helix Books, 1999.

Dan Hooper, Dark Cosmos: *In Search of Our Universe's Missing Mass and Energy*, Smithsonian Books, 2006..

Stuart Kauffman, *Reinventing the Sacred*, Basic Books, 2008.

Manjit Kumar, *Quantum: Einstein, Bohr, and the Great Debate About the Nature of Reality*, W.W. Norton & Company, 2008.

Leon M. Lederman and Christopher Hill, *Symmetry and the Beautiful Universe*, Prometheus books, 2004.

Lillian R. Lieber, *Infinity: Beyond the Beyond the Beyond*, Paul Dry Books, 2007.

David Lindley, *Uncertainty: Einstein, Heisenberg, Bohr and the Struggle for the Soul of Science*, Doubleday, 2007.

Mario Livio, *Is God a Mathematician?*, Simon & Schuster, 2009.

Thomas O. McGarity and Wendy Wagner, *Bending Science: How Special Interests Corrupt Public Health Research*, Harvard University Press, 2008.

Melanie Mitchell, *Complexity: A Guided Tour*, Oxford University Press, 2009.

Chris Mooney and Sheril Kirshenbaum, *Unscientific America: How Scientific Illiteracy Threatens Our Future*, Basic Books, 2009.

Paul J. Nahin, *The Story of the Square Root of -1: An Imaginary Tale*, Princeton University Press, 1998.

Naomi Oreskes & Erik M. Conway, *Merchants of Doubt: How a Handful of Scientists Obscured the Truth on Issues from Tobacco to Global Warming*, Bloomsbury Press, 2010.

F. David Peat, *Einstein's Moon: Bell's Theorem and the Curious Quest for Quantum Reality*, Contemporary Books, 1990.

Richard Panek, *The 4% Universe: Dark Matter, Dark Energy, and the Race to Discover the Rest of Reality*, Houghton Mifflin Harcourt, 2011.

J.C. Polkinghorne, *The Quantum World*, Penguin Books, 1986.

Alfred S. Posamentier and Ingmar Lehmann, *The (Fabulous) Fibonacci Numbers*, Prometheus Books, 2007.

Helen R. Quinn and Yossi Nir, *The Mystery of the Missing Antimatter*, Princeton University Press, 2008.

Lisa Randall, *Warped Passages: Unraveling The Mysteries of the Universe's Hidden Dimensions*, Harper Perennial, 2005.

Hilton Ratcliffe, *The Static Universe: Exploding the Myth of Cosmic Expansion*, Apeiron 2010.

Hilton Ratcliffe, *The Virtue of Heresy: Confessions of a Dissident Astronomer*, Author House, 2008.

Mark Ronan, *Symmetry Monster: One of the Greatest Quests of Mathematics*, Oxford University Press, 2006.

Ian Sample, *Massive: The Missing Particle that Sparked the Greatest Hunt in Science*, Basic Books, 2010.

Joseph Schild - Editor, *The Big Bang: A Critical Analysis*, Cosmology Science Publishers, 2011.

Donald E. Scott, *The Electric Sky: A Challenge to the Myths of Modern Astronomy*, Mikamar Publishing, 2006.

Lee Smolin, *Three Roads to Quantum Gravity*, Basic Books, 2001.

Lee Smolin, *The Trouble With Physics: The Rise of String Theory, The Fall of a Science, and What Comes Next*, Houghton Mifflin, 2006.

James D. Stein, *Cosmic Numbers: The Numbers That Define Our Universe*, Basic Books, 2011.

Paul J. Steinhardt and Neil Turok, *Endless Universe: Beyond the Big Bang*, Doubleday, 2007.

Ian Stewart, *In Pursuit of the Unknown: 17 Equations That Changed the World*, Profile Books, 2012.

Ian Stewart, *Why Beauty is Truth: A History of Symmetry*, Basic Books, 2007.

Ian Stewart, *Flatterland: Like Flatland, Only More So*, Basic Books, 2001.

Leonard Susskind, *The Black Hole War: My Battle With Stephen Hawking to Make the World Safe for Quantum Mechanics*, Little, Brown and Company, 2008.

Leonard Susskind, *The Cosmic Landscape: String Theory and the Illusion of Intelligent Design*, Back Bay Books, 2006.

Nassim Nicholas Taleb, *The Black Swan: The Impact of the Highly Improbable*, Random House, 2010.

Nassim Nicholas Taleb, *Fooled by Randomness: The Hidden Role of Chance in Life and in the Markets*, Random House, 2004.

David L. Weiner, *Reality Check: What Your Mind Knows, But Isn't Telling You*, Prometheus Books, 2005.

Peter Woit, *Not Even Wrong: The Failure of String Theory and the Search For Unity in Physical Law*, Basic Books, 2006.

-----

www.ingramcontent.com/pod-product-compliance
Lightning Source LLC
Chambersburg PA
CBHW020635220526
45464CB00001B/156